T0260119

Computers, Rigidity, and Moduli

M.B. PORTER LECTURES

RICE UNIVERSITY, DEPARTMENT OF MATHEMATICS

SALOMON BÔCHNER, FOUNDING EDITOR

RICE

H. Furstenberg, *Recurrence in Ergodic Theory and Combinatorial Number Theory* (1981)

M. Atiyah and N. Hitchin, *The Geometry and Dynamics of Magnetic Monopoles* (1988)

Yuri I. Manin, *Topics in Noncommutative Geometry* (1991)

János Kollár, *Shafarevich Maps and Automorphic Forms* (1995)

Shmuel Weinberger, *Computers, Rigidity, and Moduli: The Large-Scale Fractal Geometry of Riemannian Moduli Space*

Computers, Rigidity, and Moduli

The Large-Scale Fractal Geometry of Riemannian
Moduli Space

Shmuel Weinberger

PRINCETON UNIVERSITY PRESS

PRINCETON AND OXFORD

Copyright © 2005 by Princeton University Press
Published by Princeton University Press, 41 William Street, Princeton, New Jersey 08540
In the United Kingdom: Princeton University Press, 3 Market Place, Woodstock,
Oxfordshire OX20 1SY

Library of Congress Cataloging-in-Publication Data

Weinberger, Shmuel.
    Computers, rigidity, and moduli : the large-scale fractal geometry of Riemannian
moduli space / Shmuel Weinberger.
        p. cm.
    Includes bibliographical references and index.
    ISBN 0-691-11889-2 (acid-free paper)
        1. Riemannian manifolds. 2. Computational complexity. 3. Moduli theory.
4. Fractals. I. Title.

QA649.W45 2005
5163'73—dc22   2004044344

British Library Cataloging-in-Publication Data is available
This book has been composed in Times
Printed on acid-free paper. ∞

pup.princeton.edu

Printed in the United States of America

10 9 8 7 6 5 4 3 2 1

To   Devorah Aravah
       Baruch Hillel
and Esther Chasya Yemima

   with love and gratitude.

Contents

Preface

As I write this preface, the millenial fear of a "world-wide Y2K computer break-down" has long ago become yet another historical example of mass hysteria. However, in 1999, when I was asked to deliver the 2000 Porter lectures, computers were on everyone's mind. This book is a fairly faithful record of the lectures that ensued. The series included a popular first lecture followed by four unpopular ones.

As a result, the introduction is still semipopular, and I hope that it will garner more readers than the subsequent more technical chapters. (Most of it can be followed, I believe, by a reader who has studied multivariable calculus.) Moreover, because the techniques I am trying to explain are somewhat eclectic, there are many introductory sections scattered throughout the book (on arithmetic groups, group homology, comparison differential geometry, Kolmogorov complexity, surgery theory, etc.). This means that I sometimes end up spending more time on elementary points than on the harder theorems, and simply provide references when the going gets tough. I realize that this subjects me to accusations of violating Einstein's dictum that "one should simplify things as much as is possible, but not more than is possible." But I hope that this approach opens the ideas to a wider audience. I believe that the philosophy presented here, more than the details, has potentially broad application.

The main theme of this work is the application of the theory of computation to problems in geometry. My interest is not simply in showing the algorithmic unsolvability of natural questions, but rather in solving geometric existence problems. A second aim is to gain insight into the geometry of various moduli spaces, most notably "Riemannian moduli space," the space of isometry classes of Riemannian metrics on a given smooth compact manifold with curvature bounds. (We will see that the geometry is markedly different in low and high dimensions—with dimension four involving the most subtle computation theoretic discussion as well.)

The idea behind this is quite simple: "the method of eastern philosophy"; or less whimsically, "the logical method." It already appears in the Introduction and will be reexplained several times in the remainder of the book, at least once in each of the last three chapters, each time from a slightly different perspective.

I should explain the title. "Computers" here refers to the idealization of the digital computer, the Turing machine. Its theory will be applied to the study of the geometry of certain moduli spaces and to the existence of solutions to variational problems. Now where is the rigidity? On the one hand, of course, the very flexibility of the moduli spaces that we establish can be viewed as the opposite of too narrow a view of rigidity phenomena. More significantly, in the study of arbitrary manifolds, "eastern philosophy" requires ideas and methods that are the cornerstones of geometric rigidity (such as the theory of simplicial norms and results about hyperbolic and arithmetic groups) and topological rigidity (notably, results about versions of the Novikov and Borel conjectures, that are still known only under quite geometric hypotheses at this time).

The structure of the book is as follows. The Introduction foreshadows the main ideas of the book. The middle parts are mainly devoted to explaining and developing the necessary tools needed for the final denouement. Knowing that, in a lecture series, every lecture is someone's last, I included in each chapter some material that I think holds some noninstrumental value. The last part applies the ideas developed throughout the book to its most elaborate structure, the Riemannian moduli space. This part also has a separate, more technical, introduction. Each part ends with a section called "notes" which includes credits, historical remarks, and context that I could not include in the main text.

It is difficult to adequately express my gratitude to my collaborator, Alex Nabutovsky, who taught me most of what is novel in what follows, and codiscovered with me most of the rest. Any new results presented below are just the natural unfolding of those ideas; my main purpose here is expository. I hope there is some advantage gained in providing a connected development all in one place of our joint work, his earlier work, and their background.

Thanks are also due to Sylvain Cappell, Stanley Chang, Jeff Cheeger, Jim Davis, Benson Farb, Steve Ferry, Robin Forman, Michael Farber, Mike Freedman, Misha Gromov, Leonid Levin, Julia Knight, Misha Katz, Bob Kottwitz, John Lott, Josh Maher, Zlil Sela, Stephen Semmes, Peter Shalen, Julius Shaneson, Bob Soare, Dennis Sullivan, Bruce Williams, Kevin Whyte, Mike Wolf, and Wolfgang Ziller for helpful and encouraging conversations or correspondence, to all of the members of my Rice audience and the staff at Rice who made my stay so pleasant, and to the Courant Insitute, the Fields Institute, the University of Pennsylvania, Tel Aviv University, and the Hebrew University, who hosted me while most of this work was done.

And last of all, since there is no way to properly thank my family, Devorah, Baruch, and Esther for their multiple and respective contributions, I must be content with dedicating this book to them.

Computers, Rigidity, and Moduli

Introduction and Overview

This introduction has three goals. The first is to give some indication as to the nature of variational problems, why people study them, and what some of the methods are. The second is to explain a few of the simplest (yet most revolutionary) ideas of modern, that is, twentieth century, mathematical logic. The third is to show that these topics are not entirely unrelated.

I.1 REFLECTIONS ON LIGHT

Our story has many beginnings. One is with the ancient Greeks, who understood that a circle encloses more area than any other closed loop in the plane with the same length (or circumference). This is called the isoperimetric problem.

Probably their reasoning was inductive. First one sees that the analogous statement is true for polygons. For instance, the triangle whose area is largest with a given perimeter is the equilateral one, and similarly the square has the largest area among quadrilaterals with a given perimeter. In general, regular n-gons do best among n-gons, and furthermore, as n gets larger, they do better and better. Intuitively, the circle is the regular infinity-gon, so it must be the solution to the isoperimetric problem.

A rigorous proof of this came much later (in the nineteenth century), and even later than that came the proof of the three-dimensional analogue, namely, that the sphere is the surface that bounds the largest volume with the smallest surface area.

Such considerations can "explain" why soap bubbles in air are (fairly) spherical.[1] The volume of the bubble is determined by the amount of air trapped by the bubble, and forces on the bubble tend to try to lower its surface area.

My son and I like to blow soap bubbles, or catch one on a placid surface of water, where it forms a dome (for the exact same reason that it forms a sphere in the air) which one can then "perturb" slightly by blowing gently on it. The surface vibrates, and then the vibrations dampen and disappear, and the dome reappears. The soap bubbles one sees are in a stable equilibrium with their environment.

---

[1] Pappus was very happy to point out that the circle beats out the hexagon, and proclaimed his result an improvement over that of the bees, whom he asserted choose hexagonal tiling for beehives, which are more efficient than triangular and square tilings.

Having convinced you that my topic is at least ancient (and maybe even fun), let me begin once again, this time just about four hundred years ago, and try to convince you that the topic is "important." One of Fermat's most outstanding accomplishments, his key contribution to optics, was his explanation of why it is that, when one dips a pencil in a glass of water, it seems to bend (at the interface between the pencil and the water). His analysis is far-reaching and also explains a host of other phenomena.

Ordinarily, for example, in a vacuum (or in a nice homogeneous medium), light travels in a straight line. When we see an object, each of our eyes gives us the data about which line the object is on. Since our eyes evolved under conditions where it was reasonable to assume that light rays would be traveling in straight lines, our brain marvelously calculates the unique point of intersection of these two lines. That is where we perceive the object to be.

In an inhomogeneous medium what happens is that the light does not travel in a straight line, but moves in a more complicated way. In the case of the cup of water, the light leaves the pencil and travels in the water (which we will take to be homogeneous) and then in the air, also homogeneous. Thus, the path taken will be a broken line, that is, the union of two lines at their intersection point on the surface of the cup (which is taken realistically as having zero thickness, of course).

Of course, our eyes (really our brains) are fooled by this. We internally compute the position of each point on the part of the pencil that is in the water as if the light had arrived on a straight line from the direction that it appears to be coming from. The bend in the pencil is the result of the bend in the light. (When we completely submerge a pencil, it looks unbent, but there is still an illusion: we mistake where the pencil actually is.)

Which broken line will the light choose to traverse? The answer is not easy to predict without additional information. The effect that we are describing for water occurs with any other fluid as well, but the line will break at a different angle for different fluids.

Fermat's answer is remarkably simple: Light travels along the path that takes the least time. Light travels at different speeds in different media, and the relative speed of light in the fluid versus air will determine the path that light actually takes (figure 1).

Once you believe Fermat's explanation for a cup of water, it is but a small step to go and analyze other optical systems. For instance, in a homogeneous medium, since the shortest distance between two points is a straight line, the shortest time path is also a straight line, and therefore light travels in a straight line (as we knew before we began this whole discussion!). But one can also set up a calculation for what happens if, say, one looks at an object on the other side of an aquarium, where there are now three regions, and two interfaces, and so on. The theory of lenses and other effects is not far beyond.

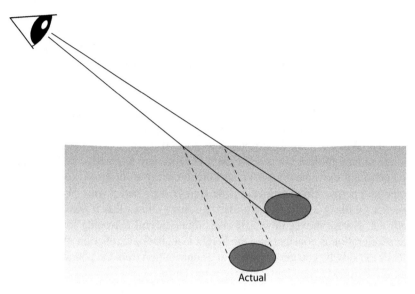

Actual

Figure 1. Fermat's principle: The perceived position is where the object would be if the light rays had gone straight.

Also, one sees (literally) the consequences of the possibility of several local minima. One sees multiple images. Here is an example of how this happens. In a desert (or on a hot paved road) the air density is lower closer to the ground, which leads to there being two paths for light to take from, say, a palm tree to the eye. There is the usual straight path from the tree to the eye, and also a second curved one that takes a "short cut" (from the point of view of time, not distance) near the ground. (Hot air is thinner, and light goes through it faster.) Our eye mistakes the second path as being the tree's reflection on the surface of some water, and we mistakenly believe that there is water nearby.

Astronomers confront this problem all the time: because of the atmosphere, the apparent positions of stars can be off from their actual positions, and all measurements have to take this into account. Last, but not least, we use this systematic refocusing of light by inhomogeneous media in the course of designing lenses, for example, in creating prescription lenses.

I.2 VARIATIONAL PROBLEMS

The form of the solution to the problem of the bending pencil is remarkable. There is a function on the space of all paths going from one point to another, and light chooses the path that minimizes some quantity that can be associated with any path (i.e., the amount of time it takes to traverse that path).

It turns out that many problems of physics have such a formulation. Euler averred that Nature always acts in such a way as to maximize or minimize some quantity. Mechanics can be very successfully reformulated in terms of action principles à la Maupertius and Hamilton. Hamilton's principle is that a system will move in such a way as to minimize the integral of the difference between kinetic and potential energies.

For instance, let us consider a falling rock. Its potential energy is proportional to its height, and its kinetic energy is proportional to the square of its velocity. When it goes from point A to point B directly below it, it must go along the straight line connecting them because any alternative path that has no horizontal movement has smaller kinetic energy at all times, and the same potential energy. (In coordinates, one always does better with $(0, y(t))$ than one does with $(x(t), y(t))$.) Similar reasoning shows that it travels along a path that always moves down. It takes more careful reasoning to prove that the acceleration is constant and the height is a quadratic function of time. If it moves too slowly, you get too much of a contribution from the potential energy, and if too quickly, the kinetic energy is too large.

This does not look like the usual form of the basic predictive question of mechanics: Given a particle at point A and its initial velocity, where does it get to after some amount of time? Instead, we figure out what path a particle must take going from A to B, and then see which of the paths corresponding to various B's have as initial velocity vector our initial v. Despite its initial peculiarity, it is hard to overestimate the value of Hamilton's principle.

(It is a pleasant mathematical fact that one can show that quite often there is a unique path of least "action" emanating from a given point A in a given direction v.)

In general relativity, a similar principle guides the trajectories of celestial bodies in the heavens. Einstein's theory describes the universe as a four-dimensional manifold with a metric (which is related to the presence of matter), and motions occur along geodesics, the analogue of straight lines.

Indeed, straight lines are themselves solutions to a variational problem. The shortest path between two points on a plane is the straight line connecting them. In fact, the Greeks defined a straight line as being the shortest path between any two of its points. (Geodesics are a natural extension of this notion to curved spaces,[2] like the surface of the Earth or of a saddle.)

(Needless to say, there are applications of geodesics that are far removed from theoretical issues of physics. Approximating Earth by a sphere, it is a matter of great significance to air travelers that the geodesics are great circles. One can prove this using the uniqueness statement I mentioned before. By rotating the sphere, if necessary, we can imagine that our points lie on the equator. For points on the equator, the equator must be a geodesic, because

---

[2]There is a minor nuance: for geodesics one usually minimizes "energy," which essentially is a device to ensure that the curves are parametrized by the arclength.

the unique geodesic connecting A to B in a given equatorial direction must be unchanged by reflection across the equator.)

Another historically important example is Dirichlet's principle, which we will discuss below. It gives a similar interpretation to the heat distribution that a homogeneous body will have if its boundary is held to a given temperature distribution. The heat distribution minimizes an "energy functional." In this example, what is minimized is not a function associated with a path, but rather a function associated with functions defined on the body ("possible temperature distributions").

In quantum mechanics the situation is a lot trickier. All of the paths from A to B enter into an integral describing the situation. What often happens is that all of the paths other than critical points end up contributing negligibly in the sense that they cancel the contributions of nearby neighbors. (This is called the method of stationary phase approximation: try estimating $\int \exp(i\lambda f(t))\, dt$ for λ large—for any closed interval not containing a critical point, it is easy to see that the integral is $O(1/\lambda)$.)

This is suggestive of another rather different reason for looking at solutions to variational problems, namely, they enter in profound analyses of the phase spaces that they are defined on. See section I.5 for a somewhat different-seeming geometric incarnation of this idea.

I.3 THE BEST IS OFTEN BEAUTIFUL

I do not know nearly enough about the history of the idea that the "best things" are always the most beautiful or most perfect. It certainly lurks behind the scenes in the solution to the isoperimetric problem. The circle is the most symmetric of all closed curves, and in at least this sense, it is the most beautiful.

On the other hand, these ideas were also behind the strong drive of the pre-Keplerians to force all celestial motions into circles.

While we no longer take this idea seriously as a scientific principle, it is invaluable as a heuristic that guides much research. The "best" way to find interesting configurations is to define a natural functional and show that extrema (exist and) are the objects sought after.

I cannot resist mentioning a problem that someone suggested to me when I was in high school.[3]

Problem

Suppose one has $2n$ points in the plane, no three on a line, n of which are blue and n of which are red. Show that it is possible to find n line segments, whose endpoints are each one red and one blue, so that the segments do not intersect.

---

[3]I am not sure who, but I believe it was the great problem lover and solver, Donald J. Newman.

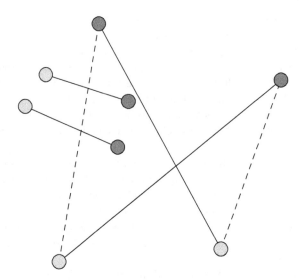

Figure 2. Connecting dots by a variational principle. Note that when we switch the connections between the dots that give intersecting segments, we lower the total length but increase the total number of intersections. The progress we make toward an intersectionless configuration is not readily apparent.

Solution

Consider the "moduli space" X consisting of all sets of n segments in the plane connecting reds to blues. (This is a set with $n!$ elements in it.) Let $L: X \to \mathbb{R}$ be the function that assigns to a configuration in X the sum of all the lengths of the arcs that constitute that configuration.

Geometric Exercise

Show that in the "L-minimizing configuration of arcs" there can be no intersections. Hint: If there is an intersection between two arcs, make a switch of which red is connected to which blue, and check that L is made smaller (figure 2).

In this case, it is clear that some configuration minimizes L; L assumes at most $n!$ values, so one of them must be the smallest! Often, showing that the function under consideration has a minimum (or maximum) value is the most difficult part. The following paragraphs show some more examples of this.

Suppose $f(x)$ is a convex function of a (positive, if you like) real variable.[4] Then $F = \sum f(x_i)$ is a convex function of the variables. If u is a minimum of

---

[4]Recall that a function is convex if its graph lies below the straight line between any two of its points. A typical example is x^2. For smooth functions, convexity follows from the positivity of the second derivative.

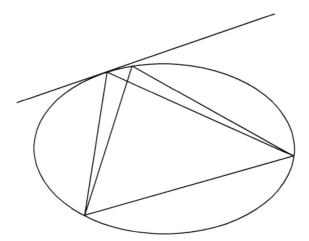

Figure 3. Billiards by a variational principle. By moving the "middle vertex" in the
direction of the excess of the angle of reflection over the angle of incidence,
we increase the circumference of the inscribed triangle.

F (perhaps subject to the constraint that $\sum x_i$ is constant), then so is σu, for any
permutation. Convexity then implies that the whole convex hull of the $\{\sigma u\}$
consists of minima, and, in particular, the vector all of whose coordinates are
the same (i.e., the vector is completely symmetric)—this common value is the
average of coordinates of u—is a minimum. For instance, setting $f(x) = x^n$,
and using n-variables, we quickly obtain the arithmetic-geometric inequality
as a special case: $(x_1 x_2 \cdots x_n)^{1/n} \leq (x_1 + x_2 + \cdots + x_n)/n$.

Without convexity, this reasoning does not work and there can be nonsymmetric minima. (Consider the function $f(x) = x$.) All that one can see is that
if u is a minimum, so is σu. Physicists refer to the phenomenon of symmetric
laws having nonsymmetric solutions as "symmetry breaking." (It is important
to note that the symmetries of physics are symmetries of the laws, not of their
solutions, i.e., not of reality. Whether this is a deep point seems to be a matter
of some controversy.) For us, we view the presence of symmetry as part of the
beauty that a variational problem is apt to have.

As yet another example, consider the functional $F = \sum |x_m - x_{m+1}|$, where
the x's are constrained to lie on a compact convex surface S; we imagine the
subscripts to be cyclically ordered; this is just a fancy of way of saying that
there is a term $|x_n - x_1|$ in the sum. Let us also insist that no two consecutive x's
coincide. One can check that the critical points of F, thought of as polygons
with n vertices on S, satisfy the condition that the angle of incidence is the
angle of reflection. (See figure 3 for the two-dimensional version.)

Morse theory (see section I.5 below) on the space of such configurations of
points on S (which is abstractly a sphere) can be used to show the existence of

"billiard trajectories" on m-dimensional tables with smooth convex boundary. (The number one gets grows "generically" like nm, where n is the number of bounces on S, and m is the dimension—we do not count reordering the hit points or reversing their order as different.) With no fancy theory, we see that there is, for each n, at least one billiard trajectory—the one that maximizes this functional. (The maximum clearly automatically satisfies the "no two consecutive equal" condition.)

And, finally, this strategy is at work in the classical ontological proof of the existence of God:

Definition.   God is the greatest being that can be imagined.

As in the above example, one studies the properties of the maximum to solve one's problem. God must exist in this view (whatever we imagined could only be greater if it actually existed), and must be omnipotent, omniscient, benevolent, and so on, indeed, the summum bonum.

We will not discuss the difficulties with this example here (although they are related to two issues discussed below, namely, Weierstrass's criticism of Dirichlet's principle (section I.6) and the use of natural language in the formulation of logical arguments (section I.7)).

I.4 MODULI SPACE (PHASE SPACE)

In the course of our previous discussion, we introduced a "moduli space" of certain configurations of arcs in the plane. It is worth taking a moment to get used to the geometric terminology. By a moduli space (sometimes called a phase space) one means a space whose points can be put in a one-to-one correspondence with the objects we are interested in. Here are some examples that the reader might enjoy verifying.

Example 1   *The space of points on a curve is just the curve itself. The space of configurations of two indistinguishable (but not necessarily distinct) points on a curve is a Möbius strip. The boundary of the Möbius strip corresponds to the locus where the two points coincide.*

Example 2   *The space of configurations of a pencil whose tip touches a specified point on the desk can be thought of as being a disk (say, centered at the point of the tip and of radius the length of the pencil). The one-to-one correspondence is just given by associating with the pencil the shadow point on the desk cast by the eraser from an overhead light. We will return to this example in the next section.*

Example 3   *The space of configurations of a short pencil (asymmetric, e.g., having a point and an eraser) whose end lies on a globe can be identified with* *the orthogonal group of* 3×3 *matrices. (The tip tells you where the first unit vector goes. The pencil then points you in the direction of the second unit vector in an orthonormal set. Finally, the third row of the matrix is determined from the first two by the right-hand rule.) If the pencil has length equal to the diameter of the globe, the moduli space is just the globe itself (there is a unique way to place the pencil, knowing where the tip is placed).*

Example 4 *The space of configurations of a short rod on the globe (the rod is entirely symmetric: the two ends are indistinguishable) is a "three-dimensional lens space with fundamental group* $\mathbb{Z}_4$*."*

All of these examples were quite low dimensional. However, higher-dimensional examples arise extremely naturally.

Example 5 *The moduli space of n points in* $\mathbb{R}^3$ *is* $\mathbb{R}^{3n}$*. If the n points are assumed distinct this becomes much more complicated. The space of noncolliding pairs of points on the line is* $\mathbb{R}^2$ *with the* $y = x$ *line removed. It is quite hard to visualize the moduli of (say) n noncolliding and indistinguishable points, especially as n gets large (e.g., for the purpose of thinking about the possible configurations of molecules of a gas in a bottle).*

Example 6 *If, in addition to the configurations in the above examples, we would like to consider their dynamics and therefore their velocities as part of our phase space, then the above spaces must be replaced by their tangent spaces.*

Example 7 *We have already had reason to think about the moduli space of all curves connecting one point to another in space (or on a general space), or of the functions on a region in* $\mathbb{R}^3$*. This gives us some natural infinite-dimensional examples.*

Example 8 *Another simple infinite-dimensional example is the space of all strictly convex bodies in some Euclidean space. This space is canonically contractible. First, any body has a center of gravity. Around this center, one can inscribe a largest ball. Then the body can be canonically deformed through convex bodies to this ball. A much more complicated space is the space of (nice) regions in Euclidean space which are diffeomorphic to the ball. Its geometry will be discussed in chapter 2.*

I.5 CALCULUS AND BEYOND

Although our interest lies in infinite-dimensional problems, it behooves us to consider first the finite-dimensional case. Let us start with a function $f : \mathbb{R} \to \mathbb{R}$ whose local minima interest us. What can we do? The standard approach of first-year calculus is to try solving the equation

$$df/dx = 0. \tag{1}$$

Solutions to (1) are called *critical points*.

Every local minimum is a critical point, but there are usually other critical points as well. First, there are local maxima. Second, there are the inflection points, whose meaning is somewhat more complicated to explain. Around these points, one might be able both to increase and decrease the value of the function, by a small movement, but not by much. (The change is of "higher order.")

The inflection points (in dimension one) are a very unstable phenomenon: they can be perturbed away. Consider, for instance, $f(x) = x^3$, which has an inflection at the origin. By "perturbing slightly," that is, adding a small term cx, where c is a small real number, we obtain the function $g(x) = x^3 + cx$ which has no inflection points. (Interestingly, for $c > 0$, there are no critical points, and for $c < 0$ there are two: a local minimum and a local maximum, which come closer together as c approaches zero, colliding at $c = 0$.)

Local minima and maxima cannot be perturbed away so easily. If we instead considered $f(x) = x^2$, which has 0 as a local minimum, then $g(x) = f(x) + cx$, no longer has the origin as a local minimum, but for all c there is a local minimum at $x = -c/2$; when c is very small, this local minimum is close to the origin.

Sometimes, we teach additional tests involving looking at the second derivative to tell us the nature of the critical point. In higher dimensions, the analogue of the test (1) for local minima and maxima is[5]

$$\text{grad}(f) = -(\partial f/\partial x, \partial f/\partial y, \ldots) = 0. \tag{2}$$

Solutions to this equation are then the critical points. (Sometimes, they are referred to as *stationary points*; we will see why in a moment.)

Even if we are looking at solutions to (2) that cannot be perturbed away, there is now another possibility, called saddle points. The prototypical example of this already occurs in two variables $f(x, y) = xy$. The origin is a critical point, but neither a local minimum nor a local maximum. Along the line $y = x$, f has the origin as a local minimum, and along the line $y = -x$, the origin is a local maximum. These facts should convince you that this critical point cannot be perturbed away.

The gradient of f, $\text{grad}(f)$, defined in (2) is a *vector field*; that is, it gives an assignment of a vector in the plane at every point of the plane. If one "pushes" a point in the direction of $\text{grad}(f)$ then one gets "flow" on the plane that usually pushes one asymptotically toward a local minimum.

---

[5]Because of my fixation with local minima, as opposed to maxima, I have defined gradients with the opposite sign to the usual convention. Of course, the definition of critical point is unaffected.

Now you see why critical points are called stationary. The "gradient flow" that pushes points around the plane does not move them. Note also that while almost all points get pushed toward the local minima, there are some (a lower-dimensional set of) points that do get pushed into the saddle points. (The dimension of this set can be determined through a second derivative test.)

Remark. It turns out that, although we made a big fuss about local minima in our discussion of variational problems, often a more careful analysis of the source of the problem shows that really the solution to the problem at hand is a stationary point rather than a local minimum. Concretely, the geodesics on the sphere are (arcs of) great circles. If one considers a great circle, say the equator, a small perturbation to, say, a constant latitude circle does lower the total length. Some people, in fact, talk about the principle of stationary action as opposed to the principle of least action, and so on, when discussing mechanics. Similarly, Fermat's principle should really be called the principle of stationary time, rather than the principle of least time, but old habits die hard.

We will get a better intuitive feel for the different sorts of critical points when we move away from flat space and move on to curved space. This is just as well, as we saw that many of the moduli spaces that interest us look like curved spaces.

The spaces we are interested in are *manifolds*. We will generically denote a manifold by M. These are spaces that locally, that is, in the neighborhood of each point, can be coordinatized. For example, while Earth is not planar, we can use latitude and longitude to describe unambiguously points anywhere except near the poles. However, even near the poles, one could make local maps that describe positions within the icecaps clearly enough.

Another example is the surface of a bagel. The Möbius strip is almost a manifold; the only problem is the boundary. If one removes the edge of the strip one has a manifold.

Typical examples of manifolds are hypersurfaces in Euclidean space. They can be described as sets of the form $g(x, y, z, \ldots) = c$ for some smooth function g.[6] (General manifolds can be embedded in Euclidean space of some higher dimension, and are "locally" the level sets of a set of functions whose derivatives are linearly independent.[7])

The sphere is the locus where $x^2 + y^2 + z^2$ is constant. (At the "poles" use (x, y) as coordinates.)

A manifold M has a tangent space TM_p at any point p. For hypersurfaces, it can be thought of as the hyperplane through that point normal to

---

[6]For this set to be a manifold, i.e., for local conditions to exist at some point, one should make the assumption that grad(g) is nonzero at the point in question. To see what can go wrong without that hypothesis, consider $g(x, y) = xy = 0$ near the origin.

[7]This is the analogue of the condition of nonvanishing grad(g) in footnote 6.

grad(g).[8] For simplicity we will say that a function defined on M is smooth (or differentiable) if it could actually be defined on a neighborhood of M in Euclidean space, and the extension is smooth in the conventional sense.[9]

If $f: M \rightarrow \mathbb{R}$ is smooth, then we can define grad(f) to be the vector field on M, that is, the assignment of a tangent vector to M at every point, by using an extension to a little neighborhood. Letting F be that extension, define

$$\text{grad}(f) = \text{the projection of grad}(F) \text{ to } TM.$$

One can see that grad(f) is well defined (i.e., independent of the extension F). For directions tangent to M, the directional derivative of f can be computed by taking the "dot product" with $-\text{grad}(f)$, all in exact analogy with the situation in flat space.

A *critical point* for f on M is a point where grad(f) $= 0$.[10] A *critical value* is the image under f of a critical point. After a small perturbation, if M is compact, we can assume there are only finitely many critical points (and, hence, finitely many critical values). Also, as in the flat case, local maxima and minima are critical points.

The key insight of *Morse theory* is that critical points and critical values are places where the topology of M changes. To understand this, I will draw the standard picture (figure 4).

Here M is a torus embedded as a surface in $\mathbb{R}^3$. Let us consider the parts of M that lie below some point p on the target real line (that is, consider the sublevel sets of f). If p is below 1, then the corresponding part of M is empty. If p lies between 1 and 2, then the corresponding part of M is (topologically) a disk. If p lies between 2 and 3, that part of M looks like a (bent) cylinder. When p lies between 3 and 4, the topology changes again, and the inverse image is "a punctured torus," and if p is above 4, the inverse image is a torus. The points 1 and 4 correspond to minima and maxima (respectively) and the points 2 and 3 correspond to saddles.

The change in topology is easiest to understand at local minima. When one crosses the corresponding critical value, the sublevel set gains another component (containing the responsible critical point). One gets a new disk.

When one goes through other critical values, the topology changes similarly. When we passed through 2, the topology changed to looking like a cylinder, which has a "one-dimensional hole," the pictured horizontal circle. Crossing through 3 also created a new "one-dimensional hole," the "inner circle." Finally,

---

[8]Note that this gradient has been assumed nonzero. As an example which shows that one cannot always find a single function, even for general hypersurfaces, the Möbius band cannot be the level set of any smooth function defined on any open set containing it.

[9]This is equivalent to the function being smooth from the point of view of all the coordinate neighborhoods of all the points.

[10]In terms of F and g, this means we are at a point where grad(F) is a multiple of grad(g); this is the classical condition of a Lagrange multiplier for finding extrema subject to a constraint.

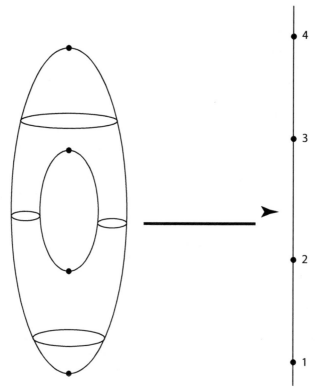

Figure 4. The classical picture of a Morse function: height on a torus. We have emphasized the critical points and the critical values.

crossing through 4 produced for us the whole torus, which is itself a "two-dimensional hole" that cannot be filled in (in itself).

The language of "holes" is the subject of homology theory. In the later chapters it will be taken for granted. For now, we will continue our discussion intuitively.

Morse theory gives a quantitative way to use measurements of the number of holes on a manifold to give lower bounds on the number of critical points of a generic smooth map. The simplest of these inequalities is that the number of local minima (and by symmetry, maxima) is at least the number of components (= zero-dimensional holes[11]).[12]

---

[11]Components can be viewed as zero-dimensional holes because by putting one point in one component and another in the other one has embedded the zero-dimensional sphere in such a way that it cannot be filled in by a one-dimensional disk, namely, an arc. Filling it in would be precisely finding a path connecting these two points, contradicting the assumption that they exist in different components.

[12]The equality of the number of zero- and n-dimensional holes in an n-dimensional compact manifold is a special case of an important relationship called Poincaré duality.

The reason for this is rather easy to see from the point of view of the gradient vector fields. If we push in the direction of the vector field, as in the flat case, almost every point ends up at a local minimum. Of course, other stationary points are exceptional.

If one does not hit a stationary point, the gradient vector field smoothly deforms one sublevel set to a lower one. Topology can change only at the critical values, when the gradient flow gets stopped at a stationary point.

The saddle points are essentially the points that correspond to higher-dimensional holes either being created or being filled in. (So the topology always does change.) In a neighborhood of the saddle points in the above picture, there are two distinguished directions: one going into the stationary point (and staying there), and a second downward direction that produces the "new one-dimensional holes," which are quite visible in the picture: they are the visible vertical circles.

Note that, coming out of the local maximum, there are two (linearly independent) directions that go downward: don't be fooled by the topmost vertical circle. It and the outer vertical circle are both part of the two-dimensional hole "caused" by the local maximum.

(More technically, for those who know homology, a generic smooth function gives rise to a chain complex, where a basis for the kth chain group is the set of stationary points with k downward directions. The homology of this chain complex is the same as the homology of the manifold.) In any case, the basic principle of Morse theory is that

> *The homology of a manifold gives a lower bound on the number of stationary points that a smooth function on it must have.*

By the way, observe that in moving to Morse theory we have geometrized the calculus technique in a way that uses the topology of the phase space to force the critical point equation (2) to have many solutions. This is rather different from our starting point, which was to try to solve that equation directly, and say that some of the solutions were local minima that interest us, or even that all of these are "stationary points" with just as significant physical meaning.

Let us apply all of this to a concrete situation. Let us consider what happens when we have a pencil with its point on a table. Everyone knows what happens: it falls down. In terms of the moduli space of such configurations (example 2 in section I.4), which we know is a two-dimensional disk, we are seeing flow along a gradient flow (of the function $1 - r^2$ where r is the distance to the origin) to the boundary, whose points correspond to the pencil lying down on the table.

But we have forgotten the local maximum! That is an unstable stationary point. (It is a critical point of index 2, as at the maximum there are two linearly independent directions to push the pencil down.) We already knew about that stationary point by thought experiment: if a pencil were put in precisely the upright position, it would not fall down. In practice, I have never seen a pencil standing on its tip when I walked into my office. Since this is an unstable stationary point, I never expect to see it; the slightest perturbation, caused, say, by my tiptoeing footsteps, will knock it down.

More interestingly, since a disk cannot be deformed into its boundary, no vector field can push points of this moduli space into its boundary. As a result, for instance, even if one has a pencil that is on a table on a moving train, which travels in any old crazy way, there will be (at least) one unstable "stationary point," which corresponds to a position where the pencil does not fall. (Of course, it would be very hard to figure out where that position is.)

Another example alluded to earlier is that we can use such reasoning to prove the existence of periodic billiard paths of any period inside a convex body, by using the L-function on the space of "cyclically ordered configurations on the body." More refined analysis of this space enables one to actually get bounds on the number of such trajectories.

Finally, I will explicitly mention that all the ideas discussed in this section can be applied to infinite-dimensional settings. The idea of critical points, that is, the vanishing of a gradient, is replaced by solutions to the "Euler-Lagrange" equation. When things work out, the whole topology of the phase space is reflected in the solutions to the Euler-Lagrange equations.

I.6 SOME FINE PRINT

In the previous section, we talked about Morse theory, which is a method of deducing that critical points exist from an analysis of the topology of the moduli space. Now it is time to talk about the fine print. Here is an example.

Example 1   *Let $M = (0, 1)$, the open unit interval. Then $f(x) = x$ is a real-valued function with no critical points.*

Of course the trouble is a lack of compactness. (Recall that a reasonable space is compact if all sequences of points in that space have a subsequence that converges to a point in the space. Here the sequence of points $1/n$—a sequence of points where the function is getting smaller and smaller—does not converge to any point in $(0, 1)$.)

This issue can also arise in the problem of finding the shortest path between two points. If one considers this on the plane with the origin deleted, there is not always a solution to the minimization problem: if the points are on opposite sides of a straight line through the origin, then there is no minimizer.

An important situation arises if $f(m) \to \infty$ as $m \to \infty$ and M is locally compact;[13] then there is at least one local minimum.[14] In other variational

---

[13]A space is locally compact if every point is contained in a compact set which includes all of the points in some neighborhood of it. This does often fail in infinite dimensions.

[14]Recall that a space is locally compact if every point has a neighborhood which is contained in a compact subset of the space. The condition $m \to \infty$ means that one has a sequence of points of M with the property that only finitely many of them lie in any given compact subset of M.

problems, it is not at all clear that the minimizing sequences will converge (not to mention what arises from saddle points). To handle this, it's necessary to have some analytic apparatus that leads to a compactness.[15] This is often accomplished by a priori inequalities.

A historically important example is the Dirichlet principle. In this case one is interested in solving the following problem that arises in the theory of heat flow:

Problem

Let Ω be a compact region in Euclidean space, whose boundary is a smooth manifold $\partial\Omega$. Let $f: \partial\Omega \to \mathbb{R}$ be a smooth function. Is there a function u on Ω, such that $u = f$ on $\partial\Omega$ and

$$\Delta u = \sum \frac{\partial^2 u}{\partial^2 x_I} = 0 ?\tag{3}$$

Certainly, everyone believed "on physical grounds" that such a solution existed. It is not very hard to prove (using the maximum principle, which is itself a consequence of the interpretation of Δu as an infinitesimal deviation of u from the average of u over a surrounding disk) that u, if it exists, must be unique.

Dirichlet's solution to this existence problem was to interpret (3) as the Euler-Lagrange equation for minimizing an "energy functional" on an appropriate function space.

Let $M = $ functions from Ω to $\mathbb{R}$ which are restricted to f on $\partial\Omega$.
We define $D(u) = \int |\mathrm{grad}(u)|^2\, dVol$.

Dirichlet pointed out that (3) is the Euler-Lagrange equation for $D(u)$. Weierstrass criticized this by pointing out that there was no real reason for the minimum of a functional like D to exist, and that it is hard to tell what kind of smoothness to expect, and so on. (Typically, spaces of smooth functions could well be incomplete in the relevant norms, certainly far from locally compact.) Hilbert ultimately rehabilitated this approach to the problem by introducing really new techniques which are now basic tools that are the property of all mathematicians: Hilbert space, functional analysis, and so on, which are partly devices for surmounting these problems.

Even after developing this new technology, which essentially changes the phase space on which we consider the question, we still have a serious obstacle. That is, is the solution that one constructs, which lives in a more complicated space, a reasonable solution to our problem? This function does not naively have derivatives, so the sense in which (3) holds is complicated. One wants smooth solutions, and, when one is lucky, one can get them, but that is another kettle of fish entirely, and, unfortunately I cannot discuss that here.

---

[15]The condition one finds in the literature that is most often used to deal with this is called the "Palais-Smale condition."

I.7 THE LIMITS OF COMPUTATION (AND OF PROOF)

Now I want to shift gears entirely to another topic, one that also has venerable roots in antiquity, but in this case, I don't think that much is lost by beginning with the insights of the twentieth century. I am talking about logic, and, for the purpose of this lecture, only of the original ideas of Gödel and Turing.

Gödel's theorem is one of the most stunning achievements of human thought, and is so universally misquoted, that like Heisenberg's uncertainty principle and Einstein's "physical relativism," it forms an essential part of the deep analytic trash talk one inevitably hears at the wrong sort of cocktail party.

We will return to Gödel in a few moments, after saying a word or two about Turing. Living as we do, in a time when computers are on every desk and in almost every pocket, it is much easier for us to apprehend the basic ideas, without getting involved in a morass of detail.

Let us consider a computer that reads instructions written in some language, say, English. We could convert every English instruction into a number by the following artifice:

$$A \rightarrow 2$$
$$B \rightarrow 3$$
$$C \rightarrow 5$$

and so on, assigning to each letter a prime number, and then continuing through all the punctuation marks. This uses up the first 30 primes (say). Then start over again and assign to A the 31st prime (and the 61st, and 91st, and so on) and to B the 32nd one (and the 62nd, the 92nd, and so on), C the 33rd, and on and on.

Given a sentence or even a book, one can now multiply together all the prime numbers associated with all the letters and punctuation marks and get a (very big) number. The first letter will be assigned a prime from the first time the letters are listed, the second letter/space/punctuation takes a prime from the second list, and so on.

Not every number comes from a sentence: 1 and 4 don't because there is no combination of letters that produces these numbers, and many other numbers do not come up because they arise from gibberish (nonwords) or ungrammatical constructions.

Still, an English speaker will (by definition) be able to recognize exactly which numbers correspond to words, sentences, paragraphs, commands, well-reasoned logical arguments, and poetry.[16] This is called the *arithmeticization* of English.

By the way, English is not really a good language for our purposes, nor is any natural language. We have trouble with ambiguous words and sentences. The

---

[16]I got carried away. We do not need poetry for our discussion.

point of "computer languages" is that, at least with regard to texts instructing us to do things, one can make a language where each grammatical sentence can be unambiguously decoded.

Part of the technical work in Gödel's proof of the incompleteness of arithmetic was making sure that arithmetic could be written in such a fashion. Turing did the same thing with his machines and their arithmeticization.

If we wanted to, we could go even further and list all the integers that correspond to sentences, arguments, or the instructions of a computer program in numerical order. For instance, the smallest words in our ordering of English are "a," "I," and "ah," which corresponded to 2, 23, and $2 * (77 = $ 39th prime), but are now 1, 2, and 3, respectively.

In short, we can now discuss notions like the "first word," "second sentence," and, what is really important in the following, the "nth program" (or the "mth proof"). For convenience we will deal only with computer programs that operate on (that is, accept as inputs) positive integers. Arithmeticization, as above, lets such machines do things like word processing, control factories, and interact with humans in video games. Let us consider a few computer programs and nonprograms.

Program 1:

> Input an integer k.
> Add one to k, and call this k.
> Output k.

This program just adds one to an integer.

Program 2:

> 1. Input an integer k.
> 2. Add one to k and call this k.
> 3. If the result of line 2 is less than 100 go back to line 2.
> 4. Output the answer.

This program is a bit more complicated, and I found it necessary to number the lines. What it does takes a little longer to decode. If you input an integer less than 100 it outputs 100, and from 100 on, it adds one.

Here is a last one (for now).

Program 3:

> 1. Input an integer k.
> 2. Multiply a by 3 and call the result k.
> 3. Go back to line 2.

This program is quite loopy! For every integer the machine just sits there and never outputs anything. (I would say that it is not doing anything, which is right for all practical purposes, but, from its own point of view, it is busy working and working.) Many other programs would be just as useless as program 3 (for instance, if line 2 were replaced by line 2 of program 2). We assert that program 3 does not halt for any input i.

Now consider

Program 2'

1. Input an integer k.
2. Add one to k and call the result k.
3. If the result of line 2 is more than 100 go back to line 2.
4. Output the answer.

This program halts for $k < 100$, and does not halt for $k > 99$.

Finally, we come to Turing's theorem:

Theorem (Turing) *There is no program (a synonym is algorithm) that will decide in general whether program a will halt on input b.*

To be pedantic, there is no program that will output 0 on input $2^a 3^b$ if program a halts on input b, and will output 1 on $2^a 3^b$ if program a does not halt on input b. (We do not care what the program would do to integers divisible by other primes.) We will get to the proof in a moment, but it is worth seeing that this implies

Gödel's Theorem

Within any finite axiom system there are arithmetic propositions about the integers that can neither be proved nor disproved.

Without being too precise about what is meant by a finite axiom system, note that, with an infinite system, I could (with the help of a clairvoyant) list all truths about the integers, and then just look to see whether my proposition is on the list! Proof here need not necessarily be what we normally think of as "mathematical proof"; it just has to be something which

1. works (i.e., proved statements are in fact "true"),
2. can be checked (i.e., if I write down a proof, everyone else (e.g., any proof checking computer) should agree with me that the proof is valid), and
3. has proofs that can be written down as a series of "English" statements.

The second condition can be paraphrased as saying that "proof" should not be a matter of opinion. In fact, we suppose that there is an algorithm that decides whether or not a series of sentences comprises a correct proof.

The third condition (i.e., "English") means that there is a "good" language that expresses the proof—not a natural language.

Mathematical proof has these properties. (Some other types of "reasoning" could also be approached by similar techniques.)

Gödel's theorem is *not* asserting that "any sufficiently complicated system cannot understand itself" or anything like that. People like to say things like that as a prelude to dismissing some neuropsychological study as being, a priori, impossible. I think we would be very happy to understand ourselves to the point of understanding the mechanisms that make us do what we do, even if we could not always calculate or predict what these mechanisms would cause us to do in some particular circumstance.

However, what Gödel's theorem does say is absolutely astounding. It tells us that a completely well-defined mathematical system, like the positive integers, cannot be completely analyzed by mathematical tools. The axioms might determine the system up to isomorphism, yet still not be strong enough to enable us, by the method of proof, to determine all the facts we would like to know about the system.

Gödel's theorem was first proved by encoding a classical Greek paradox (the liar's paradox) into the proof system, and has subsequently been proved by encoding various other paradoxes. However, the Turing proof I am about to sketch is somewhat more concrete. An example of the kind of statement that cannot be proved or disproved (in general) is that "program a does not halt on input b."

For some a's and b's this can be proved or disproved. We gave examples above. But for every proof system in the sense described above, there will be a program a which does not stop on input b, but for which no proof of this fact exists. (I hope it is clear that for every program c which does stop on program d, there is a proof that c stops on d—at least with "conventional proof" as our proof system.)

The proof of Gödel's theorem from Turing's is conceptually quite simple. It's the old story of having a room full of monkeys typing Shakespeare.

Suppose that for every (a, b) there were a proof that a either halts on b or does not. Let us define a program of the sort that Turing ruled out:

1. Input a and b.
2. Let $k = 0$.
3. Replace k by $k + 1$.
4. Check if "proof k" is a proof that a halts on b; if so, output 0 and stop.
5. Check if "proof k" is a proof that a does not halt on b; if so, output 1 and stop.
6. Go back to step 3.

Under our assumption that Gödel's theorem is false, there will be some $I > 0$ so that "proof I" works in either line 4 or line 5. (Here we used properties 3 and 2 of our requirements of proofs: property 3, so that we could talk about "proof k," and property 2, so that we could check what the proof is proving!)

Now, since condition 1 is correct, we will have decided whether or not a really halts on b. QED

The proof of Turing's theorem is not very hard either. Suppose we had a program P that decided the halting problem. Now let us consider a new program:

Program T

1. Input k.
2. Invoke P to decide whether program k halts on input k.
3. If k does not halt on input k, output 1 and stop.
4. If k does halt on input k, compute the effect of program k on input k, add 1, output this number and stop.

Program T is a program assuming that P is. Therefore, it can be written out in excruciating detail and then assigned a number t. Now, I ask, what does T do on input t? Certainly it halts; T halts on all inputs, so the program will stop at line 4, but the output will be, by definition, $T(t) + 1$. Since $T(t) = T(t) + 1$ is impossible, P could not in fact have existed. QED

This argument is a form of Cantor's diagonal argument which he used in the nineteenth century to prove that there are different sorts of "infinities" (indeed an infinite number of different kinds). It is remarkable that the same argument that grants us the power to begin analyzing the infinite also tells us that there are limits about what we could hope to know about finite integers!

As a bit of terminology, we will henceforth refer to programs as *Turing machines* or simply as *machines*.

I.8 AND BEYOND

The ideas of Gödel and Turing are so revolutionary that they immediately raise a host of new questions. In this chapter I will not be able to pursue them, but it seems like a good idea to raise them here to set an agenda and develop some terminology.

Gödel's theorem says that there are undecidable questions, that is, propositions which cannot either be proven or disproven. But are there natural ones? Are there problems that people cared about, where they were undecided not out of lack of imagination, but rather because there was no solution?

Gödel himself gave such an example: he showed that it was impossible to prove the consistency of arithmetic (assuming it was indeed consistent).

Turing's approach gives another example: that a machine will never halt on a given input is sometimes undecidable. Debugging computer programs and proving that they calculate some specific function are (in general) "impossible" tasks.

In fact, such questions arise naturally in set theory, in point set topology, in infinite group theory, in combinatorics, and elsewhere. We will discuss some of these in some detail, especially in the next chapter. Let us introduce some terminology.

Definition. A function $f : \mathbb{N} \to \mathbb{N}$ is *computable* if there is a Turing machine that computes f. (Note: Here our function f need not be everywhere defined. Where the machine does not halt, f is undefined. To make the terminology consistent with that of the rest of mathematics, it would be best to call such a thing a "partial function," but we shan't.) A set S in $\mathbb{N}$ is *computable* if its characteristic function is computable.[17] Completely equivalently, it is computable if there is an algorithm which determines whether or not a natural number lies in S.

Remark. Later we will have to enlarge our scope and consider other types of sets, for example, sets of propositions or of manifolds or functions, and discuss their computability. This will always be done by means of an arithmeticization process.

A weaker condition on a set S is that of being *computably enumerable* (c.e.).[18] S is c.e. if there is a computable function whose range is S, or, equivalently, the partial function which is 1 on S and undefined elsewhere is computable. S is computable if and only if (iff) both S and its complement are c.e.

Using arithmeticization, we note that the sets of provable propositions (about integers) and of machines that halt on the input 17 (for instance) are both examples of c.e. sets that are not computable.

Notation

$K = \{2^a 3^b$: such that machine a halts on input $b\}$.

Note that if somehow we had access to K, we could find out whether any proposition could be proved: we would build a machine to try to prove proposition b. It would halt iff there is a proof of b.

Making this idea of "having access to K" precise is done through the device of an "oracle." Let us imprecisely think of any set S as defining an oracle, by just imagining that we can go to Delphi and ask whether or not an integer lies in S. Oracles can certainly enable us to solve problems we could not have solved before, and we are interested in what can be computed by machines that have access to an S-oracle: these will be called the S-computable functions.

---

[17] Recall that the characteristic function of a set is the function that assigns 1 to elements of the set and 0 to the nonelements of the set.

[18] We are following the suggestion of Soare that one emphasize the computational aspects of this notion; in most of the literature, this notion is called *recursively enumerable* (r.e.) and *computable sets* are called *recursive*.

Exercise

If one has access to an oracle for K, show that it is possible to decide membership in any c.e. set.

So, in some sense K is the hardest c.e. set. There are other sets that are more difficult to decide membership in than K, but they are not c.e. An example is

$$\{\, n \mid \text{machine } n \text{ halts for infinitely many inputs}\}.$$

But we will confine our attention here to c.e. sets.

Exercise

If S is a computable set, then the S-computable functions (and sets) are exactly the computable functions (and sets).

Now we are ready for

Post's Problem

Is it the case that for any c.e. set A the A-computable functions are either computable or the same as the K-computable functions?

The answer turns out to be negative (as Post expected), as was proven by Friedberg and Muchnik (independently). This was a watershed moment in "computability theory."

Let me elaborate a bit on what Post's problem asks, by introducing some terminology. Let us say $A < B$ (read A is *reducible* to B) if one can compute (the characteristic function of) A using a B-oracle. This is a partial order ($A < B$ and $B < C$ implies $A < C$, and always $A < A$). To say A is reducible to B means that (in a very crude and fundamental sense) A is at least as easy to compute as is B.

This notion is significant because that is often how undecidability or noncomputability propagates. We showed that there were undecidable propositions, because if not, we could solve the halting problem. Almost all the undecidability results we referred to above, and that we will use later, come via reduction to some noncomputable set of integers, say (and typically) K. The idea is to encode in some way a noncomputable set S in the other discipline, so that something you care about in the other discipline holds iff the encoding integer lies in S. We will return to this in detail in the next lecture.

There is another way to look at Post's problem, via the "busy beaver function."

Definition. Let $BB(n)$ be the supremum of the number of steps of computation of the computation of machine a on input b, where we consider only $a < n$, $b < n$, such that a program actually halts on input b.

$BB(n)$ grows very quickly.

Proposition *For any computable function f everywhere defined, $BB(n) > f(n)$ for all sufficiently large n.*

The function f can be crazy from any conventional point of view. A relatively slowly growing function of this sort would be $g(n) = \exp(\exp(\cdots(n)))$ where there is a tower of n exponentials, and $f(n) = g(g(\cdots(n)))$, where we have a stack of $g(n)$ g's. It is somewhat daunting to try to understand what $f(3)$ is.

Without loss of generality f can be assumed increasing. Consider the machine that on input n computes $f(n)$, then goes through some internal looping for that number of times before stopping and outputting 1. This machine computes the constant 1, but its halting time is larger than $f(n)$ (for every n). Whenever n is larger than the Gödel-Turing number of this machine, the inequality in the proposition will hold.

For any c.e. set, we can consider the *stopping function* for a Turing machine that recognizes that set—it is the maximum number of steps taken by that Turing machine to stop on any of the elements of the set in the interval from 1 to n; BB is essentially the stopping function for K. The stopping function of a noncomputable c.e. set can never be bounded by a computable function. Indeed, a c.e. set is not computable iff the stopping time of its defining Turing machine is not (bounded by) a computable function. The same considerations show that BB cannot be bounded by the stopping time of any set used in the solution to Post's problem, although that function is itself not bounded by any computable function.

Later we will return to these considerations from an "applied" point of view. But, for now, I would like to point out a line of thought about scientific explanation that leads directly to considerations of relative computation.[19]

The goal of scientific theories is to explain nature, or perhaps better, predict nature. How is this done? What would satisfy us? Concretely, let us imagine that a bee's flight satisfies some complicated nonlinear partial differential equation. It seems quite conceivable that no method of calculating solutions to that equation could be completed in polynomial time. Would this bother us?

Practically, it certainly would. We like to make predictions before events happen, and here we would not be able to.

On the other hand, we do not really think that bees solve nonlinear equations or that electrons "obey" Schrödinger's equation. What we think is that they satisfy equations. Quite conceivably the description of nature will involve producing theories whose consequences are extremely hard to figure out, even

---

[19] I would like to acknowledge the influence of ideas of Manin, of Bar-Av and Nabutovsky, of Geroch and Hartle, and of Traub on the following discussion and its follow-up in the last chapter.

in principle. Conceivably, the result of some experiment will be undecidable within the theory (thought of as an axiom system).

How could a theory like that be verified? I believe quite easily. To be quite concrete, suppose a term like $\mu = \sum 2^{-s}$, where the sum is over the elements S of a noncomputable set, came up. Say S was the set K. That would be terrific. By doing some very refined experiments, we could find out whether or not an integer lies in K.[20]

We would learn, say, that machine 3 never halts on input 50. Every whir of the machine would be another test for the theory. We would learn that it does halt on input 51, and someone would apply for a National Science Foundation grant to test this prediction. We would be very impressed by the verification when this occurred after two centuries of computation.

In fact, this shows that certain scientific theories could well lead to analog computers that can do computations that digital computers, that is, Turing machines, cannot. That is a wonderful possibility that dwarfs even the hopes held out by quantum computing, which would "only" lead to much more rapid computations of functions that were, in any case, computable.

However, the theory could well require computations of new quantities in terms of things like μ (after all, all sorts of "constants of nature" enter in complicated ways into formulas for other observable quantities). Moreover, once we believed the theory, and believed that μ really is what we say it is, then we could use it as the basis for computations solving many undecidable problems. I emphasize that even without a theoretical description of μ, we will still be doing calculations relative to a μ-oracle. Thus, the theory of "physical" computations is exactly the subject of computability relative to a set.

Very interestingly, some models of quantum gravity seem to involve numbers like our μ. Some of the issues they raise are intimately connected to the considerations of these lectures.

I.9 THE METHOD OF EASTERN PHILOSOPHY

Armed with our understanding of computability and undecidability, let us return to geometry and, in particular, to Morse theory.

There are some situations where Morse theory guarantees very little. For instance, if M is connected and compact, and $f : M \to \mathbb{R}$ is smooth, we are guaranteed nothing more than a single local minimum in general. If M is noncompact, we are not even entitled to one, even if we insist that the function be bounded from below, without the condition that the function is *proper*.[21]

---

[20]Leonid Levin has argued that the scenario I suggested about the motion of a bee does not actually survive critical examination. However, I do not believe that his arguments lead to the conclusion that the constants of nature must be computable real numbers nor do they preclude the existence of a theoretical interpretation of these constants. They would call in doubt the epistemological foundations of such a theory.

We shall see that there are circumstances related to undecidability that force there to be infinitely many local minima.

If the idea of Morse theory is that to solve an Euler-Lagrange type equation, one wants to study the topology of phase space, that is, the aggregate of all of the system's possible states, then we may describe our method by the following, slightly whimsical, but useful maxim, which we call

The Method of Eastern Philosophy

To understand yourself, understand what those who could have been you could have been.[22]

In other words, it is not enough to study the phase space associated with your problem. One can gain information by considering the phase space of "doppelgangers," that is, "those that could have been you." What do we mean by a doppelganger? Either the doppelganger is you, or it is not, and if it is not, how could it have been you?

We shall not give a precise meaning to this term; like hard, tall, deep, and intelligent, it is a term that carries within it an implicit relativism. (I might seem tall, but wait until Goliath enters the room.) With this proviso, a doppelganger is a system that is "very hard" to distinguish from the one I'm interested in studying.

One way to make this precise would be to have an infinite set of systems, one for each natural number. Imagine that the set of integers for which the relevant system is "equivalent" to mine is not computable. Then, no doubt, this set of systems must contain many doppelgangers (whatever they are, and however we choose to measure "hard").[23]

Let us make matters concrete. Consider the space of knots, that is, embeddings of one sphere in another two dimensions higher, $\text{Emb}(S^n \subset S^{n+2})$. It turns out that this space has infinitely many components, in other words, there are infinitely many knots that cannot be deformed (isotoped) into one another. That is the traditional study of knot theory. We are interested in one component, the space of unknots. (It is the component of the "equator of the equator.")

It turns out that for $n = 1$, that is, knots in ordinary space, there is an algorithm to determine whether knots are isotopic, but for $n > 2$ that is not the case. (We will prove this in a later chapter.) What does this mean? Consider some special class of knots, say, ones described as polyhedra or as the images

---

[21] A real-valued function is proper if it goes to infinity for any sequence in the space that goes to infinity.

[22] I shall not explain the sense of "eastern" in this description.

[23] Unless we allow ourselves to perform tasks that are computably difficult like computing a busy beaver function.

of polynomial functions,[24] with, if one insists, rational coefficients. Then one can arithmeticize this set, and ask which of these knots are isotopic to the trivial knot. One can even build a machine to try to build an isotopy; the program will halt on a given input iff the knot is isotopic to the unknot. Thus, the unknots form a c.e. set.

In terms of the hierarchy we discussed before, this set is equivalent to K, the general halting set. The method of proof is by "encoding." For each c.e. set S, one produces a series of very special knots K_r for each natural number r, in such a way that r lies in S iff K_r is isotopic to the unknot. (The K_r have easily computable descriptions as polynomials or as polyhedra.)

Now suppose F is a functional, defined on the whole space of embeddings, which we will suppose is very nice, is computable, and has a computable gradient flow.[25] We are trying to show that F has infinitely many local minima on the component of the unknot.

Suppose, by way of contradiction, that F as a functional on the component of the unknot has only one local minimum.[26] We assume nothing about how many critical points F has on other components.

We will use this information to produce an algorithm to distinguish the unknot from the other knots. We can think about what we are doing in terms of the geometry of a cactus patch. If one stares at a cactus patch, it very hard to tell whether or not two points that are a few feet apart really lie on the same cactus plant or lie on neighbors (see figure 5). *Consequently*, each individual plant has to have complicated geometry.

It is this word "consequently" that we need to explain.

Suppose the opposite is true, that is, suppose each individual cactus plant has simple (Morse) geometry; then imagine raindrops rolling down plants in the patch that touch these two points we are interested in. With extremely high probability, these raindrops will collect at local minimum points. If some plant P had a unique local minimum, we would be able to use the rolling raindrops to test whether an arbitrary point lies on P, and whether an arbitrary point could be connected to the ones on P.

In our situation, one uses the gradient flow associated with F in place of gravity pulling on raindrops, and one sees that the component of the unknot cannot have a unique local minimum.

But the exact same argument proves that there are infinitely many local minima. For, if we suppose not, then one could list them, say, $m_1, m_2, \ldots, m_k$.

---

[24] The Tarski-Seidenberg theorem tells you that one can tell whether a polynomial function defined on the sphere is an embedding (see the notes at the end of the chapter).

[25] Alas, there are any number of technical details that must be suppressed. I beg the indulgence of the reader and ask for some suspension of disbelief.

[26] If F is invariant under some group of diffeomorphisms, we shall assume that there is a unique critical orbit.

Figure 5. A cactus patch.

(Any finite set can be listed, although there might not be an algorithm to produce the list.[27]) One flows down the cactus (I mean the space of unknots) and looks to see whether one arrived at any of, say, $m_1, m_2, \ldots, m_k$. If so, then one's starting point was on P, and if our list of local minima was comprehensive, the converse holds as well. Thus, we still get a contradiction.

As this book goes on, we will discuss various functionals on various spaces for which this argument works, more quantitative forms of it, variants, and so on. In fact, we will begin going down that road in the very next section, but will not give substantive details until the later parts of this book.

I.10 FRACTALS AND GEOMETRICIZATION

I would like now to suggest another point of view on "eastern philosophy," or "the logical method," which is through the lens of "dichotomy theorems."

A dichotomy theorem has the following general form: Not only can't my moduli space be disentangled from all the others (computably), that is, I can't

---

[27]We are not giving the algorithm here; we are showing the existence of an algorithm. We do not have to produce an algorithm to get a contradiction, because there is no algorithm at all that will recognize K.

tell whether or not a knot is actually knotted, it can't be disentangled from those components that have some property X (which the original moduli space typically does not have, or is not known to have), for example, ones whose complexity (measured in some fashion) is large.

Such theorems are proven by the method of encoding, but now the encoding has to be done more carefully to ensure that the doppelgangers have property X.

In Hilbert's tenth problem, for instance, the main result is that one cannot tell whether or not a diophantine equation has any solutions; a dichotomy version of this result would assert that there are classes of diophantine equations for which one knows a priori that they have either no solutions or exactly one, but it remains undecidable whether or not equations of this class have solutions. The extra information that there is no more than one solution was of no help.

This is useful in a few ways. One is somewhat technical: the property X could be a compactness (or Palais-Smale) type condition. In that case, even though our original F on our original moduli space might not satisfy this condition, for the purposes of applying some Morse-theoretic arguments, one can pretend that the compactness condition does hold.

Another way this is useful is in learning about the geometry of these moduli spaces. Suppose that one can prove that the moduli space P we are interested in is simply connected,[28] but our dichotomy theorem asserts that the doppelgangers must be nonsimply connected. Then P must have big regions containing relatively small loops which do not bound disks until one moves very far away (or allows the areas of the disks to be very large). Otherwise, one would use the geometry (reflected in the topology of large regions) of the moduli space to distinguish the doppelgangers from the "real thing."

This ultimately tends to imply that there are a great many different sorts of regions within the moduli space, in each of which the moduli space looks quite different.

I should emphasize that the type of geometry we are discussing is "large scale." Let us make the following definition.

Definition. The *depth* of a local minimum m for F—at length, or scale, s—is infinite if m is a global minimum. If not, then there is a sequence of points $m_1, m_2, \ldots, m_k$, so that $m = m_1$, $d(m_n, m_{n+1}) < s$, $F(m_k) < F(m)$. Then depth = inf sup$(F(m_n) - F(m_1))$. In other words, the depth is a measure of how high you *must* go on your way to lowering the value of the functional (ignoring some amount of local fluctuation).

Usually we are not interested only in fixed lengths and depths, but it is often convenient to tie these to the value of the functional. For instance, we might choose to be interested only in exponential depth local minima, which would mean that we would be interested only in local minima that are of depth

---

[28] This means that all closed curves can be contracted within the space.

$> \exp(F(m))$. We might choose to let the length scale also be $\exp(\exp(F(m)))$, so these guys are local minima for quite a distance around.

Addendum

When the method works, it produces infinitely many $G(F)$-deep local minima, for any fixed computable function G.

The $G(F)$-sized balls around these local minima tend to be quite different regions in moduli space. The local minima are providing us with a useful vehicle for making explorations of the geometry of moduli space.

In the last part of this book, we shall develop these ideas further and make them more quantitative. For instance, we will use hierarchies of functions to be able to deal with many different ("large scale") scales and shall see that not only are there very deep local minima, but they are also "spaced together somewhat closely." The deep ones are spaced much further apart than the more shallow ones.

We show this by applying the same technique that produces the local minima more globally to make them rather dense; that is, we produce doppelgangers from each point on the moduli space, rather than just once for the space. (This requires a slightly more complicated encoding.) Of course, the deeper the local minima the further apart they have to be.

From where do we get the freedom to produce families of local minima and to vary their depth? From the choice of which Turing machine to encode into the (enlarged) moduli space (containing the doppelganger components). The "stopping function" turns out to control both the density of the local minima, and the depths of the minima constructed.[29]

The picture is somewhat reminiscent of Koch's fractal snowflake. Here one starts with a triangle and replaces the straight lines by broken lines, repeatedly. The process and the result of the first few steps are shown in figure 6.

The first step is an equilateral triangle, but the second replaces it by a "Star of David" which accounts for the 60° rotational symmetry one sees subsequently (figure 7).

This picture (with an inverted choice of scales, which get larger and larger) suggests that there is some sort of "large-scale fractal structure" to spaces that have suitable logical complexity. Of course, the different portions of moduli space are much more different from each other than the different parts of a snowflake. This argues against another aspect of fractality, self-similarity.

With this warning, I still like to use (at least tentatively) the expression "large-scale fractal" in discussing these spaces. It is somehow an evocative description

---

[29] In fact, stopping time functions end up to be more suitable for our purposes than the c.e. degrees. In the last chapter, we will, in fact, even get minima using computable sets—albeit ones given in a hard-to-compute way.

The basic step in creating "snowflakes":
Replace a staight line above
by a broken line as below.

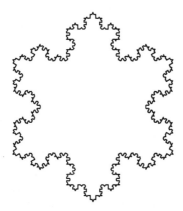

Figure 6. The basic step in the creation of a snowflake.

Figure 7. A snowflake.

of the type of structure of moduli that we are seeing. The reason for this is that many of the ideas in variational theory are based on the ideas of Morse theory and calculus on manifolds, with its assumptions of linearity at a small scale. On the other hand, on a snowflake the height function has many more critical points than are forced by its topology (topologically, it is just a circle). If one picks a threshold to pay attention to, and considers only local minima of at least that depth, the smaller the threshold, the more local minima one sees and the denser they are in the snowflake.

What I would like to suggest then is the general principle that analysis on moduli spaces for problems that have logical complexity resembles in some way what one would expect of differential calculus on fractal-like sets.

The rest of this book will be devoted to making precise statements based on this principle and developing the method of eastern philosophy into a rigorous tool to verify them.

NOTES

Below are references for most of the topics discussed in this introduction. I have not given references to popular books and articles, although for some of these topics, there are some excellent choices. I tried to pick books that really get to the core of the matter and that have either style or efficiency or some other good feature to recommend them. Needless to say, this bibliography is antiencyclopedic.

Some useful references for variational principles in elementary physics are

V. Arnold. *Mathematical Methods of Classical Mechanics*. Springer-Verlag, New York, 1989.

R. Abraham and J. Marsden. *Foundations of Mechanics*, 2nd ed. Revised and enlarged. With the assistance of Tudor Racedlatiu and Richard Cushman. Benjamin/Cummings, Advanced Book Program, Reading, Mass., 1978.

Y. Choquet-Bruhat, C. Dewitt-Morette, and M. Dilliard-Bleick. *Analysis, Manifolds, and Physics*. 2 vol. North-Holland, Amsterdam, 1982 and 1989.

The three-volume Feynman *Lectures in Physics* (Addison-Wesley Longman, Reading, Mass., 1980) are an amazing read, as is his popular book, *QED: The Strange Theory of Light and Matter*, published by Princeton University Press, Princeton, N.J., 1985.

A good book on mathematical methods for studying variational problems is

M. Struwe. *Variational Methods: Applications to Nonlinear Partial Differential Equations and Hamiltonian Systems*, 3rd ed., Springer-Verlag, Berlin, 2000.

Most introductory books on differential geometry explain the variational theory of geodesics, which we will return to below. Some also explain the theory of minimal surfaces, Plateau's problem, or the theory of harmonic maps, which are other significant chapters in the theory of variational problems.

The classic on Morse theory is

J. Milnor. *Morse Theory*. Princeton University Press, Princeton, N.J., 1963.

Robin Forman has written a very nice account of some of the key ideas of Morse theory for an audience just familiar with multivariable calculus. It can be obtained from his web page: http://www.math.rice.edu/~forman. (Besides that, one can also find there nondifferentiable versions of the theory very suitable for combinatorial applications, and applications to combinatorics.)

There are very many sources for homology theory. Two very different ones are

R. Bott and L. Tu. *Differential Forms in Algebraic Topology*. Graduate Texts in Mathematics 82. Springer-Verlag, New York, 1982.

J. Vick. *Homology Theory: An Introduction to Algebraic Topology*. 2nd ed. Graduate Texts in Mathematics 145. Springer-Verlag, New York, 1994.

The connection of billiard trajectories to the perimeter functional on the space of "inscribed n-gons" is due to Birkhoff, who showed the existence of periodic trajectories with n bounces for every n in the plane. The extension of this idea using Morse theory on "cyclic configuration spaces" is due to M. Farber and S. Tabachnikov, and is developed in several papers. The first one is

M. Farber and S. Tabachnikov. *Topology of cyclic configuration spaces and periodic trajectories of multi-dimensional billiards.* Topology 41 (2002), 553–589.

For Dirichlet's principle and a good gentle introduction to partial differential equations, I personally like

G. Folland. *Introduction to Partial Differential Equations.* Princeton University Press, Princeton, N.J., 1976.

For logic, I recommend

M. Davis. *Computability and Unsolvability.* McGraw-Hill Series in Information Processing and Computers. McGraw-Hill, New York, 1958.

J. Schoenfield. *Mathematical Logic.* Addison-Wesley, Reading, Mass., 1967.

Y. Manin. *A Course in Mathematical Logic.* Translated from the Russian. Springer-Verlag, New York, 1977.

Davis is considerably more elementary than the others, and concentrates a bit more on computation. For computability theory, there are the original and modern classics

H. Rogers. *Theory of Recursive Functions and Effective Computability*, 2nd ed. MIT Press, Cambridge, Mass., 1987.

R. Soare. *Recursively Enumerable Sets and Degrees: A Study of Computable Functions and Computably Generated Sets.* Perspectives in Mathematical Logic. Springer-Verlag, Berlin, 1987. (New edition to appear.)

The philosophical remarks concerning the connection between computation and physics were spurred on by conversations with the following authors and by reading their papers:

Y. Manin. *Mathematics and Physics.* Translated from the Russian. Progress in Physics 3. Birkhäuser, Boston, Mass., 1981.

A. Nabutovsky and R. Bar-Av. *Non-computability arising in dynamical triangulation model of four-dimensional quantum gravity.* Commun. Math. Physics 157 (1993), 93–98.

R. Geroch and J. Hartle. *Computability and physical theories.* Found. Phys. 16 (1986), no. 6, 533–550.

J. Traub and A. G. Werschulz. *Complexity and Information: Lezioni Lincee.* Cambridge University Press, Cambridge, 1998.

A. Nabutovsky. *Logic phenomena of Euclidean Quantum Gravity.* Preprint available at http://www.math.toronto.edu/nabutovsky/

The "method of eastern philosophy" will be developed further in this book, but under the more conventional name "the logical method." A view of its development can be found in

A. Nabutovsky. *Non-recursive functions, knots "with thick ropes and self-clenching "thick" hyperspheres.* Commun. Pure Appl. Math. 48 (1995), 381–428.

————. *Geometry of the space of triangulations of a compact manifold,* Commun. Math. Phys. 181 (1996), 303–330.

A. Nabutovsky. *Einstein structures: Existence versus uniqueness*, Geom. Funct. Anal. 5 (1) (1995), 76–91.

A. Nabutovsky and S. Weinberger. *Variational problems for Riemannian functionals and arithmetic groups*, Publ. Math. d'IHES 92 (2000), 5–62.

————. *The fractal nature of Riem/Diff I*. Geometriae Dedicata 101 (2003), 1–54.

The last of these references is also the source for the material in the very last section. Finally, the undecidability of high-dimensional knot theory is quite simple and can be found in

A. Nabutovsky and W. Weinberger. *Algorithmic unsolvability of the triviality problem for multidimensional knots*, Comment. Math. Helv. 71 (1996), no. 3, 426–434.

We will discuss related points in chapter 2.

In the footnotes we mentioned the Tarski-Seidenberg theorem. It is one of the earliest and most significant applications of logic to geometry. It shows that Euclidean geometry is a decidable theory. There are a number of different versions of this theorem; let me mention two.

Definition. A real *semialgebraic set* is a subset of $\mathbb{R}^n$ that is defined by a system of polynomial identities, nonidentities, and inequalities.

Tarski-Seidenberg Theorem (Geometric Form)

If one projects a real semialgebraic set from $\mathbb{R}^n$ to $\mathbb{R}^m$, then its image is also a semialgebraic set.

Notice that if one just started with identities (i.e., algebraic sets), the projections would require inequalities. For instance, the projection of $y = x^2$ to the y axis is the nonnegative reals, which is not algebraic.

This theorem is exactly equivalent to a logical statement:

Tarski-Seidenberg Theorem (Logical Form)

The theory of the real numbers with $+, *, 0, 1, >$ is decidable; indeed, it admits elimination of quantifiers.

The statement about Euclidean geometry I mentioned follows by encoding geometry into algebra, that is, the process that we call vector analysis or analytic geometry.

What "quantifier elimination" means is the following. Propositions essentially have the form that, for all x, there is a y such that for every z you can find a w so that if something or other involving all these variables holds, then something else does as well. Quantifier elimination says that any such sentence is equivalent to a sentence not involving all the "for all"s and the "there exist"s.

Note that giving a "there exists" statement is almost immediately the same thing as projecting down a less complicated (logically) set involving extra variables. Thus, the equivalence between these statements is not hard to see.

The logical form of the theorem makes us realize just how versatile semialgebraic sets are. They are exactly the sets that are definable in the language of real numbers and its algebraic operations.

In terms of the citation in the text, the question of whether a polynomial is 1-1 on the sphere is clearly a semialgebraic set (of polynomials of a given degree, i.e., thought of as described by the coefficients). It is $\{ f \mid$ for all x, and for all y, if x and y lie on the sphere and $f(x) = f(y)$ then $x = y \}$.

Later on we will need facts like the (effective) triangulability of semialgebraic sets, or the decidability of whether a semialgebraic set is connected. They can be found in

R. Benedetti and J. Risler. *Real Algebraic and Semi-Algebraic Sets*. Actualités Mathématiques Hermann, Paris, 1990.

J. Bochnak, M. Coste, and M. Roy. *Real Algebraic Geometry*. Ergebnisse der Mathematik und ihrer Grenzgebiete (3) 36. Springer-Verlag, Berlin, 1998.

L. van den Dries. *Tame Topology and o-Minimal Structures*. London Mathematical Society Lecture Note Series 248. Cambridge University Press, Cambridge, 1998.

Chapter One

Group Theory

It is not surprising, given the syntactical nature of arguments in logic and in combinatorial group theory, that one of the earliest examples of a natural, widely studied, mathematical problem to be shown algorithmically unsolvable lies in group theory. Our unsolvability (and dichotomy) results in topology and geometry will come about by encoding group theory into these subjects, which will be done in chapter 2. The goal of this chapter is to provide the necessary background in group theory.

This subject contains a number of deep theorems. Some of the highlights of this part are:

- the unsolvability of the word problem by Novikov and Boone, and the triviality problem by Adian and Rabin (section 1.2);
- Higman's embedding theorem (section 1.2);
- the recent work of Sapir with Birget and Rips on Dehn functions (section 1.3);
- the results of Borel-Wallach and of Clozel on the cohomology of arithmetic groups (section 1.5);
- the theorems of Baumslag and Dyer with Heller and Miller on the group homology of finitely presented groups (section 1.6).

1.1 PRESENTATIONS OF GROUPS

Let G be a group. G is said to be *finitely generated* if there are finitely many elements $g_1, g_2, \ldots, g_k$ such that every element of G is a product of these elements (many times) and their inverses. All finitely generated groups are countable, but the rational numbers give an example of a nonfinitely generated countable group.

Note that saying that G is finitely generated is exactly the same thing as saying that there is a surjection from a free group $F_k \to G$ for some k. We say that a subgroup H of G is *finitely normally generated* if there is a finite set S such that H is the smallest normal subgroup of G containing S. Alternatively, the elements of H are products of things of the form gsg^{-1}, where s is from S, and g lies in G.

The group G is finitely presented if there is a surjection $F_k \to G$ whose kernel is finitely normally generated. A set R which normally generates the kernel is called a set of relations for G. One uses the notation

$$G = \langle g_1, g_2, \dots, g_k \mid r_1, r_2, \dots, r_s \rangle,$$

where the r's are a list of the elements of R. The r's should be thought of as being words in the g-letters. One can easily show that the kernel of one surjection of a free group to G is finitely normally generated iff it is for any other surjection. Using this, one can show, for instance, that the following group is not finitely presented:

$$\langle x, y \mid [x, x^a y x^{-a}] \text{ where } a = 0, 1, 2, 3, \dots \rangle = \mathbb{Z} \wr \mathbb{Z},$$

where $[g, h] = ghg^{-1}h^{-1}$ is the commutator of g and h, and $\wr$ denotes the wreath product (for those familiar with this concept; we will never have any use for it).

One thinks of the finite presentation as giving the group defined by these generators, subject to the given set of relations (like an axiomatic system). The word problem asks one to give an algorithm for deciding whether a combination of generators represents the trivial element of G; in other words, whether a certain potential relation is a consequence of the relations that are already part of the set R. (Again, this is a property of a group, not of the way the group is presented.) We will discuss the word problem in the next section.

Remark. It is not at all easy to tell if two finite presentations define the same group. The Tietze theorem asserts that two presentations define the same group iff there is a sequence of elementary moves (and their inverses) that relate the two presentations. The first move adds a new generator and a new relation that defines the generator in terms of the old ones, for example,

$$\langle x, y \mid \,\rangle = \langle x, y, z \mid z = xyx^{-215}y^{12}x \rangle.$$

(We use the convention that a relation of the form $r = s$ is just a user friendly way of writing rs^{-1}.) The second move allows one to add a relation that is a "consequence" (i.e., that lies in the normal closure) of the others. So

$$\langle x, y \mid yx = xy \rangle = \langle x, y \mid yx = xy, xyxy = yxyx \rangle.$$

These moves always increase the complexity of presentations, but that means that their inverses can decrease complexity. We will see that relating two "simple presentations" by Tietze moves could involve going through very complicated intermediate presentations. (We will never really need the Tietze theorem, but it will sometimes be useful for illustration purposes.)

Besides their natural logical interest, finitely presented groups occur extremely naturally throughout mathematics. Probably the simplest, general sources of these are fundamental groups of compact manifolds and arithmetic groups (or more general lattices in real Lie groups).

The reason that compact smooth manifolds have finitely presented fundamental groups is because they are all homeomorphic to finite polyhedra (= finite simplicial complex) by a relatively difficult theorem of Cairns and Whitehead (CW), or by Morse theory, which, at least, shows that they are homotopy equivalent to finite CW complexes.

Given a finite CW complex or simplicial complex one can very simply write down a presentation of the fundamental group. The 1-skeleton is always homotopy equivalent to a wedge of circles; the 2-cells of the 2-skeleton then give the relations. The proof of this description goes by way of van Kampen's theorem, which describes the fundamental group of the union of two spaces that intersect "nicely." To describe the answer, we shall need the constructions of "amalgamated free product" and "HNN extension."

Definition. Let A, B, and C be groups, and let C be a subgroup of both A and B. $A *_C B$ is the group defined to have the universal property that A and B both map to $A *_C B$, and if $f : A \to D$ and $g : B \to D$ are homomorphisms that agree on C, then there is a unique extension of f and g to a map $A *_C B \to D$.

General nonsense implies that $A *_C B$ is unique (up to canonical isomorphism) if it exists. To construct it, we can give a formula. Suppose $A = \langle a_1, a_2, \ldots \mid r_1, r_2, \ldots \rangle$, $B = \langle b_1, b_2, \ldots \mid s_1, s_2, \ldots \rangle$, and C is generated by $c_1, c_2, \ldots$. As C is a subgroup of both A and B, we can write $c_1 = c_1(a\text{'s})$ and $c_1 = c_1(b\text{'s})$, $c_2 = c_2(a\text{'s})$ and $c_2 = c_2(b\text{'s})$, etc. Now

$$A *_C B = \langle a_1, a_2, \ldots, b_1, b_2, \ldots \mid r_1, r_2, \ldots, s_1, s_2, \ldots, c_1(a\text{'s})$$
$$= c_1(b\text{'s}),\ c_2(a\text{'s}) = c_2(b\text{'s}), \ldots \rangle.$$

(Note that actually one needs only a homomorphism from C into both A and B for the definition and universal property to make sense. The universal property is helpful for a conceptual understanding of why the amalgamated free product is independent of all the choices involved.) With this preparation, we can state van Kampen's theorem.

Van Kampen's theorem

If two spaces with fundamental groups A and B intersect along a connected space with a fundamental group C, then their union has the fundamental group $A *_C B$.

The HNN extension is a natural analogue of the amalgamated free product, and comes up in determining the fundamental group of a union when the intersection is not connected.

Definition. Let A be a group and B_j, $j = 1, 2$, isomorphic subgroups (let ϕ be the isomorphism). Then the HNN extension A^*_B is defined by the universal property that, if $f : A \rightarrow D$ is a homomorphism which restricts to conjugate maps on the two copies of B, then it extends uniquely to A^*_B. In terms of generators and relations, the formula is

$$A^*_B = \langle a_1, a_2, \ldots, t \mid r_1, r_2, tb_1t^{-1} = \phi(b_1), tb_2t^{-1} = \phi(b_2) \ldots \rangle$$

with the obvious notation.

If one glues a cylinder $Y \times I$ whose fundamental group is B along both of its ends to a space with fundamental group A, then one gets into a situation where the HNN extension is defined. Indeed, the fundamental group of this union is the HNN extension.

Exercise

Show that van Kampen's theorem together with the above addendum for cylinders together suffice to deal with all unions of polyhedra along subpolyhedra.

It is worth noting that these constructions can behave quite oddly when the "subgroups" are really groups with homomorphisms with nontrivial kernels. For instance, if we consider $\mathbb{Z}_2$ and $\mathbb{Z}_3$ amalgamated "along" $\mathbb{Z}$ which maps surjectively to both groups, one obtains the trivial group.

On the other hand, if all of the "inclusions" are really injections, then A automatically injects into $A *_C B$ and into A^*_B. In fact, there is quite a natural normal form for elements in the amalgamated free product and HNN extension, given a choice of coset representatives for the subgroup. In the next section we will make extensive use of these constructions, and, in particular, this injectivity statement.

The proof of this and several of the other basic theorems of combinatorial group theory can be written down in a completely opaque combinatorial fashion, but are actually quite transparent from a geometric point of view.

Definition. A graph of groups is a graph Γ such that each vertex v is associated a group G_v and each edge is assigned G_e. For each of the two inclusions of endpoints u, v in an edge e, there are given injections $G_e \rightarrow G_u$ and $G_e \rightarrow G_v$.

Notice that an edge can have both endpoints being the same vertex, in which case one has a group with two isomorphic subgroups. Thus, the data for an edge are exactly the data necessary for defining amalgamated free products or HNN extensions. For any connected graph, one can then inductively define, using amalgamated free products or HNN extensions on individual vertices, the fundamental group of the graph of groups. If all the vertex and edge groups are trivial, this is just the fundamental group of the graph.

One could define this notion in a less ad hoc way by an appropriate universal property; we leave this for the reader. One can also see that it is the fundamental group of a union of spaces, which overlap only in pairs, and that "fit together" according to the pattern of the graph (vertices corresponding to spaces, and edges to overlaps) and with the fundamental groups of every piece determined by the labels on the graph.

The following proposition gives a connection between group actions on trees and graphs of groups.

Proposition   *If a group G acts simplicially on a tree T without inversions (i.e., invariant edges are fixed), then the quotient T/G is a graph of groups, with fundamental group G, wherewith each vertex or edge is associated its stabilizer. Conversely, every graph of groups comes from an action of its fundamental group.*

The proof is little more than covering space theory. Note that, as a consequence, A injects into $A *_C B$, since the latter group acts on a tree, with A and B as vertex groups and C as the edge group. Similarly, A injects into the HNN extension A^*_B because the latter also acts on a tree with A as a stabilizer subgroup for a vertex. Here are some other corollaries:

1. Subgroups of free groups are free. (Freeness is equivalent to having a free action on a tree; free actions restricted to subgroups remain free.)
2. Exercise: Show that any subgroup of finite index in a nonabelian free group is a free group of higher rank. Show that the commutator subgroup of a nonabelian free group is a free group of infinite rank. What are its generators?
3. Generalized Kurosh subgroup theorem: Any subgroup of a graph of groups is a graph of groups where the edge and vertex groups are subgroups of the original vertex and edge groups.

The usual Kurosh theorem corresponds to free products, that is, where edge groups are trivial, so for the subgroup all edge groups are also trivial, so that the fundamental group is a free product of subgroups of the free factors and free groups.

Exercise

Supposing that G and H are nontrivial groups, not both $\mathbb{Z}_2$, show that $G * H$ contains a free group of rank 2.

1.2 PROBLEMS ABOUT GROUPS

Now let us return to the theory of group presentations. Recall that the word problem for a finitely presented group is to determine when a word represents the trivial element.

Example 1   *Finite groups have a solvable word problem. (Use their multiplication tables.)*

So do finitely presented residually finite groups, that is, groups G with enough homomorphisms to finite groups to catch any nontrivial element. (In other words, each nontrivial element g of G is mapped nontrivially by some homomorphism of G into a finite group.) The proof of this goes as follows. Start two machines going. The first lists elements of the normal closure of R systematically (i.e., going though the elements of G to get many conjugates of the elements of R, and taking many products of these) and checks to see if the given word occurs. If this happens, the machine yells "w is trivial." The second machine lists all homomorphisms from G into any finite group (why is this algorithmic?) and then checks if the word is mapped trivially. If it is not, this machine yells "w is nontrivial." Clearly only one of these will happen, and if G is residually finite, one of them will.

One could want to know a bound on how long it would take an algorithm to determine if w is trivial. In general, it could be quite bad.

Another example is the free group, where the algorithm is quite simple; one just looks for appearances of symbols like xx^{-1} within the word, and removes these. When one is done, one has a reduced word, and every group element has only one expression as a reduced word. This is an extremely fast algorithm. (It is linear in the length of the word.) There is a large and rich class of groups with a linear time solution to the word problem (subgroups of products of hyperbolic groups in the sense of Gromov; see the references), but I do not know of many general theorems for them.

Nowadays, there are even examples of finitely presented solvable groups with unsolvable word problems! But this is running ahead in our story. The wonderful theorem of Boone and (P. S.) Novikov is simply the following:

Theorem   *There are finitely presented groups with an unsolvable word problem.*

We will not even sketch the proof here. The overall idea is to encode a Turing machine into a finite presentation (using a series of amalgamated free products and HNN extensions) so "the normal form theorem" (for elements of an amalgamated free product or HNN extension) implies that the only way that a word will be trivialized is via the appropriate computation of the Turing machine. That, the construction, and the proof that it works, will give you the theorem. A number of refinements and extensions will be of importance to us later. We will get to these.

A rather different approach to these problems comes about via the following landmark result:

Higman Embedding Theorem

A finitely generated group is a subgroup of a finitely presented group iff it has a computably enumerable set of relations.

(For infinitely generated groups with a computable set of generators, the same is correct.) The necessity of the computable enumerability of the relation set is easy. The generators of the subgroup are some specific words, and all their relations are specific relations that hold in the original group. We can always enumrate the relations in a finitely presented group, by taking all products of conjugates of the relators.

By now there are a number of different proofs of Higman's theorem. We will give some references below. The theorem is remarkable in that it relates a basic group theoretic notion, embeddability in a finitely presented group, to a computation theoretic one. Moreover, it very quickly gives rise to a proof of the Novikov-Boone theorem, as follows. Let S be a c.e. set which is not computable. Consider the finitely generated, computably presented group

$$G = \langle a, b, c, d \mid a^k b a^{-k} = c^k d c^{-k} \text{ for } k \in S \rangle.$$

Note that G is a free product with amalgamation of two free groups $\langle a, b \rangle$ and $\langle c, d \rangle$. The relation $a^k b a^{-k} = c^k d c^{-k}$ is true iff k is an element of S, by the normal form theorem for amalgamated free products. Thus we cannot tell in G whether a word represents the trivial element.

Embedding G into a finitely presented group gives a finitely presented group with unsolvable word problem. By the way, Higman's technique shows the existence of a "universal" finitely presented group, which contains all others. Later we will make use of such groups.

Now let us turn to the triviality problem, which was solved by Adian and Rabin in much greater generality.

Definition. A *Markov property* of a group is a property such that (1) if G has this property, so does any subgroup of G, and (2) there is some group H not possessing this property.

Theorem (Adian and Rubin) *There is no algorithm to decide if a finite presentation has any particular Markov property.*

So one cannot tell if a group is trivial, if it is finite, abelian, nilpotent, solvable, free,[30] has a solvable word problem, is torsion-free, or contains infinitely divisible elements.

---

[30]Exercise: Show that it is impossible to decide whether a group is freely generated by a specific set of elements. Hint: Use HNN extensions.

We shall give the proof of this, since it is quite simple, given what we know about amalgamated free products and because the method is very important.

Proof. Given a finitely presented (f.p.) group G and an element w of G, we shall construct a new f.p. group G_w such that (1) either w was the trivial element, in which case G_w is the trivial group, or (2) w is nontrivial, in which case G_w contains G as a proper subgroup.

Note that this immediately gives the impossibility of deciding triviality, since an algorithm for this would give an algorithm for deciding whether w is trivial. It also implies the theorem in general, by picking G to be a free product $H * K$, where K has an unsolvable word problem, and H does not have the given Markov property. The group $(H * K)_h$ has the Markov property iff h is the trivial element, which cannot be discerned algorithmically.

Let $G = \langle x_1, x_2, \ldots, x_k \mid R \rangle$ and let w be an element of G. We form $G * F_2$, where F_2 is generated by t, s.

The normal form theorem for elements in a free product gives us a free group of rank $k + 1$ in this free product, and we will choose one generated by $w[w, t]$, and $x_{a'}^{\alpha}$, where $\alpha_a = s^a t^a$. We use the standard notation that $[\ ,\ ]$ denotes the commutator and "exponentiation indicates conjugation" in groups. The condition on the α_a ensures that the words $x_{a'}^{\alpha}$ have little cancellation possible among them, and therefore generate a free group. (This is a formal matter given the normal form theorem for free products, and we leave an exact construction as a worthwhile exercise for the reader.) One can add on conjugates of s and t by complicated words and still have a free subgroup of $G * F_2$.

Let A be a group containing a free subgroup F on $k + 3$ generators, the first one of which normally generates A. (Many of the fundamental groups of knot complements in the 3-sphere have this property. The meridian[31] of the knot is always an element which normally generates the group; the fundamental group of the Seifert surface is always a free group, and if we omit one of them, then the remaining ones together with the meridian will also generate a free group; within this free group one can increase the rank at will.) Set $G_w = (G * F_2) *_F A$.

Clearly, if w is nontrivial, this free product is nontrivial. If w is trivial, then A dies because w trivial kills $w[w, t]$, which has been identified with a normal generator of A. Once A dies, so do all the elements of F, but these each go to conjugates of the remaining generators of $G * F_2$, and thus this whole group is killed as well, completing the verification of the construction.

Finally, armed with this construction, one easily builds groups $H * G_w$ which have a given Markov property if and only if $w \neq e$.

Appendix: Some Refinements and Extensions

The study of the algorithmic problems about groups did not end with the unsolvability of the word problem; indeed, that was just a beginning.

---

[31] This is the small circle that bounds a tiny 2-disk which intersects the knot once.

To understand the first extension, we need the notion of Turing reducibility. Let S and T be sets of natural numbers. We will say that $S \leq T$ if one can compute S from an oracle that decides membership in T.

Notice that, even if T is c.e., S need not be; for instance, the complement of T is always computable from T. However, we shall restrict our attention to c.e. sets. We shall say that S and T *have the same (c.e.) degree*[32] if $S \leq T$ and $T \leq S$; in other words, if each can be computed in terms of the other. Note that we shall identify c.e. sets with the Turing machines that define them (say, as being the set of inputs on which the machine halts).

Now, the set of words representing the trivial element is a c.e. set, and therefore represents a c.e. degree. After learning that it is possible for this set to be noncomputable, it becomes natural to ask whether there are any restrictions on the c.e. degrees. The following theorem completely answers this question:

Theorem (Fridman, Clapham, and Boone) *If D is a c.e. degree, then there is a finitely presented group with degree exactly D. More precisely, there is a uniform construction starting from a Turing machine T producing a group $G(T)$, whose word problem is of the same degree as (the halting set of) T.*

One way to prove this is to give a precise version of Higman's embedding theorem which preserves the c.e. degree of the word problem. Then the construction described above certainly would yield this theorem.

There are other natural problems about elements of a f.p. group that these theorems imply are algorithmically undecidable, and of arbitrary degree, as one varies the group.

Theorem *There is no algorithm to decide, in general, whether a group element*

1. *is of finite order,*
2. *is of the form $[x, y]$,*
3. *lies in the center,*
4. *commutes with another given element, or*
5. *is conjugate to another given element.*

The proof for (5) is obvious; conjugacy is a more general problem than triviality. (3) and (4) can be proven simultaneously by considering a group $\mathbb{Z} * G$, for G a group with an unsolvable word problem. The only element that commutes with tg (t in $\mathbb{Z}$, g nontrivial in G) is the identity, which is the whole center, so the word problem for G reduces to either of these problems. To prove (1) it would suffice to observe that one can produce groups as above which are torsion-free, which is true. (They can be built up from the trivial

---

[32]These are sometimes referred to as Turing degrees, or as degrees of unsolvability.

group by HNN extensions and amalgamated free products; as the reader can check, this implies that the group is torsion free.[33])

Another approach would be to produce a recursively presented group and then Higman embed. So let

$$G = \langle a_1, a_2, \ldots \mid a_k^{f(k)} \rangle,$$

where the exponent $f(k)$ is 0 if the kth Turing machine does not halt on input k and is the number of steps that machine takes to halt, if it does. This G clearly has unsolvable "torsionality," and it can be embedded into a f.p. group.

Only (2) requires (as far as I can see) more trickery. Rather than prove it, let me just point out that it is a close cousin to the word problem and the conjugacy problem (5). If one considers a space with fundamental group G, the word problem asks one to decide whether a curve bounds a disk; the conjugacy problem asks whether a pair of curves bound an annulus. (2) asks whether a curve is the boundary of a punctured torus, whereas (4) asks whether two curves could be homotoped to lie on torus.

In this context it is worth observing that the question of whether a curve is the boundary of some compact surface is solvable. This is the same as asking whether it is a product of some number of elementary commutators, i.e., whether the element is trivial in the abelianization of the group.

Remark. The conjugacy problem is clearly "harder" than the word problem. In fact, Clapham showed that for an arbitrary pair of c.e. degrees $D \leq E$, there is a finitely presented group with word problem of degree D and conjugacy problem of degree E.

While we have mentioned only briefly (in the Introduction) that there are degrees besides computable and K, the degree of the halting problem, actually, the structure of the set of degrees, is extremely rich and complicated. In particular, it is a dense partial ordering, and there are also many noncomparable degrees, and so on.

We will close this section with the comment that for arithmetic groups, the word problem and conjugacy problems are in fact solvable (although it would be nice to get good bounds). However, the following generalized word problem is not solvable:

Theorem (Mikhailova) *There is no algorithm in $F_2 \times F_2$ to decide whether a given element lies in the group generated by a given finite set of elements.*

Since $F_2 \times F_2$ is a subgroup of $\mathrm{SL}_n(\mathbb{Z})$ for $n > 3$, we obtain the unsolvability of the generalized word problem for these arithmetic groups.

---

[33]Notice that the groups produced by HNN and amalgamated free products starting from the trivial group all have finite-dimensional Eilenberg-MacLane complexes, which gives a "geometric proof" of the nonexistence of nontrivial torsion.

The beautiful construction is irresistible. Let H be a two-generator group with an unsolvable word problem. (Such exist; even without this fact, one could modify the construction slightly, as the reader can readily see.) Write $H = \langle x, y \mid R \rangle$. Now consider the subgroup S of $F_2 \times F_2$ generated by (x, x), (y, y), and $(1, r)$ for r in R. It is trivial to check that the pair (u, v) lies in S iff u and v represent the same element of H.

1.3 DEHN FUNCTIONS

The Dehn function of a presentation of a group is a concrete measure of how hard it is to solve the word problem. More precisely, we ask, for all words of length $\leq n$ that actually represent the trivial element, what is the largest number of relations it is necessary to use (when a relation is used twice, it is counted twice) in order to prove this.

This function does depend on the presentation. However, its order of growth does not. More precisely, the Dehn functions of two different presentations satisfy a relation of the sort

$$g(n/B) - Cn - D < f(n) < g(Bn) + Cn + D \tag{1}$$

for some constants B, C, D. (The linear term is there for fairly trivial reasons: see example 1 below.)

The famous Gromov hyperbolic groups are those whose Dehn functions grow linearly. Remarkably, if the Dehn function is subquadratic, it is automatically linear. Thus, there is a nontrivial subject of the possible Dehn functions of f.p. groups.

In this section, I would like to explain a little bit about their theory and describe some remarkable work that gives an almost complete solution to the problem of characterizing Dehn functions by M. Sapir with J. C. Birget and E. Rips.

It is important to be clear that Dehn functions measure the difficulty of a particular method of trying to solve the word problem; there could be other solutions that are much more rapid. They are the analogues of the stopping times of particular Turing machines defining a given c.e. set. However, unlike the Turing machine situation, for a given group, the Dehn function has a much stronger well-definedness property (1).

Although "eastern philosophy" as we introduced it in the Introduction is phrased in terms of arbitrary algorithms, in chapters 3 and 4, we will give versions that can make use of Dehn functions, rather than arbitrary stopping times for arbitrary solutions of the word problem. Let us compute some examples.

Example 1   *The Dehn function of $\mathbb{Z}$. Let $\mathbb{Z} = \langle x \mid \; \rangle$. Then the Dehn function is trivial. However, for the presentation $\langle x, y \mid y \rangle$, the Dehn function is linear. One removes all y's one at a time, and the number of y's present can be linear in the word length.*

Exercise

Verify the linearity of the Dehn function for a free group.

Example 2 *The Dehn function of $\mathbb{Z}^2$ is quadratic. Consider $\mathbb{Z}^2 = \langle x, y \,|\, [x, y] \rangle$. Now, we know that in $\mathbb{Z}^2$ we have $[x^n, y^n] = e$. However the "proof" of this is quite large: $[x^n, y^n] \sim [x, y]^{n^2}$ (see figure 8, where "$\sim$" means that on the right-hand side we have a product of that number of conjugates of $[x, y]$). It is not very hard to see, by hand, that there is no smaller product of (conjugates of) relations that gives $[x^n, y^n]$, so one has that $D(4n) \geq n^2$. It is also easy enough to see that a quadratic number of uses of the relations suffices to "reduce" any word to normal form.*

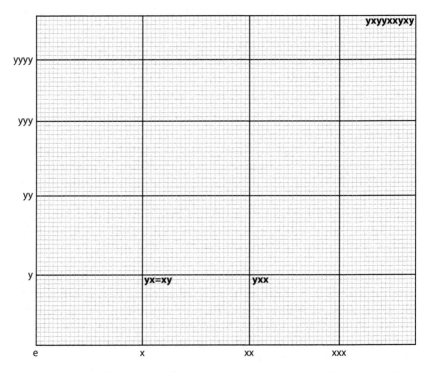

Figure 8. The Cayley graph of Z^2. (Vertices are labeled by group elements; we have labeled them using words that describe specific paths from e to the given node.) Note that the "curve" bounded by a large square is the boundary of quite a lot of small squares. The small unit squares represent the basic relator.

Note, by the way, that using the two homomorphisms $\mathbb{Z}^2 \rightarrow \mathbb{Z}$ (killing x and y, respectively), we can get a linear solution to the word problem for $\mathbb{Z}^2$. I am not sure whether there are any reasonable results about the class of groups that have a linear time solution to their word problems. One can often very substantially compress Dehn functions using "nonsyntactic" algorithms. Before going further, it is worth pointing out a limit to this:

Theorem *The Dehn function of a f.g. group is computable iff the word problem is solvable.*

If the Dehn function is computable, then one knows how many relators to try to combine to trivialize a word; conversely, given a solution to the word problem, one can then search out products of relations for all the words of length n that are, in fact, trivial. When this has been accomplished, then look at the length of the longest one you found. This is a computable upper bound for the Dehn function. Then check all products with a smaller number of relations to make sure you did not miss the shortest product realizing these words, to actually compute the Dehn function.

Example 3 *The Heisenberg group H of 3×3 upper diagonal unipotent integer matrices has a presentation with two generators and two relations:*

$$H = \langle x, y \mid [x, [x, y]], [y, [x, y]] \rangle,$$

and has a cubic Dehn function. We have $[x^n, [x^n, y^n]] = e$ in H, but it is a product of a cubic number of relators.

The way to see this is to think somewhat geometrically about the meaning of Dehn functions. Let Z be the universal cover of a space (say a manifold, to be specific) with fundamental group π. Words that represent the trivial element are essentially the same thing as closed nullhomotopic loops in the base, and, by covering space theory, these lift to closed loops in Z. Since they are nullhomotopic, these loops bound 2-disks. The Dehn function is equivalent to seeking the smallest area disk in Z that can be found to bound an arbitrary closed curve of length L.

In examples 1 and 2, we were discussing the Euclidean line and plane, respectively, where the answers are linear and quadratic. (By the way, note that as the dimension of the Euclidean space increases, the Dehn function remains quadratic.) To do example 3, consider the map $f : H(\mathbb{R}) \to \mathbb{R}^2$ (of real 3×3 unipotent upper diagonal matrices) to $\mathbb{R}^2$ (giving the entries immediately above the diagonal) by killing the commutator. Observe that $f^*(dx \wedge dy) \leq dA$ (the area 2-form). Since $dx \wedge dy$ is a closed 2-form, one can integrate it over any disk bounding a given closed curve, like the one represented by $[x^n, [x^n, y^n]]$, and (1) the integral is independent of the bounding disk (even surface) and (2) by the area inequality, it gives a lower bound on the area of such a bounding disk. Playing with these gives the cubic nature of the Heisenberg group.

Remark. This cubic nature is due to the identity $[x^n, y^n] = [x, y]^{n^2}$, so we have n^2 commutators to commute with x^n. This identity asserts that the copy of the integers generated by $[x, y]$ in H is *quadratically distorted*. Using this fact, one can also see that the number of distinct elements of the group represented by words of length n (the *volume growth* of the group) is quartic.

Remark.   A finitely generated subgroup of $F_2 \times F_2$ has nonrecursive distortion exactly when the generalized word problem associated with that subgroup is unsolvable.

Example 4   *Let us now give some exponential and superexponential examples; with a bit of trickery, one can promote these higher up the Ackermann hierarchy.*[34]

The solvable Baumslag-Solitar group $G = \langle x, y \mid xyx^{-1} = y^2 \rangle$ is actually a linear group. Then $[y, x^n y x^{-n}]$ is a linear length word that requires an exponential number of relations to kill. Note that $x^n y x^{-n} = y^{2^n}$. Now, let us add a new generator as follows:

$$\langle x, y, t \mid xyx^{-1} = y^2, tyt^{-1} = y, txt^{-1} = x^2 \rangle.$$

Then $[y, [t^n x t^{-n} y, t^{-n} x^{-1} t^n]]$ is a linear length word that now needs 2^{2^n} relations to kill, and so on.

We leave the verification to the reader. A useful method for doing these types of calculations is to try to consider "van Kampen diagrams," which are directed planar graphs, where every edge is labeled by a generator of the group, and every face is labeled by a relator of the group. A word w is trivial iff there is a van Kampen diagram over the given presentation, whose boundary is that word.

In all of our examples, the relations are designed for easy examination of possible van Kampen diagrams. We shall not pursue this direction any further,[35] but now move on to some very general theorems.

Theorem   *(Birget-Sapir) If $D(n)$ is the Dehn function of a finitely presented group π, then it is the stopping time of some nondeterministic Turing machine that solves the word problem for π.*

Here is the idea of this restriction. Suppose we knew that (a function equivalent to) n^α were a bound on the Dehn function of G. Then I would check 2^{n^α} products of relations to see which words were trivialized, and this would give me $D(n)$ exactly. Thus, if the Dehn function were actually (up to equivalence) of the form n^β, we would be able to compute β quite quickly. It turns out that, when you unravel this, it implies that the first k digits of β can be computed in

---

[34]The Ackerman hierarchy measures how much induction is used in the definition of a function. nm is adding m to itself n times; iterating that, we have m^n, which is multiplying m by itself n times; then one can consider $m^{[n]}$, which would exponentiate itself n times. After that, it becomes more awkward: $m^{[[n]]}$ would be doing $m^{[m]}$ to itself n times, and so on. The Ackermann function is the diagonalization of this $A(2) = 2^2$, $A(3) = 3^{[3]}$,....

[35]It seems quite possible that the extension of some of our geometric results to dimension four will depend precisely on the continuation of these explicit methods.

time $2^{2^{2^k}}$. In other words, if the Dehn function grows quite slowly, it cannot take too long to compute it. The best result known currently is almost the converse of this observation.

Theorem   *If $T^4(n)$ is superadditive ($T^4(m+n) \geq T^4(m) + T^4(n)$), and T is the stopping function for a (perhaps nondeterministic[36]) Turing machine, then $T^4(n)$ is equivalent to the Dehn function of a finitely presented group π.*

The group is built out of the machine whose stopping time is T. More precisely, if M is this machine, there is an injective map from the input words of M to the words of p such that word size does not get distorted under this map, and the word is accepted by M iff its image is trivial in π.

The moral is that, besides the extra fourth power and the assumed superadditivity, which seem like technical conditions, one gets a remarkably close connection between stopping times of general Turing machines and Dehn functions.

1.4 GROUP HOMOLOGY

Group homology and cohomology is a very basic tool that developed simultaneously with homological algebra. My impression is that early researchers were very highly motivated by Hopf's result that the cokernel of the two-dimensional Hurewicz homomorphism depends only on the fundamental group of the space. Nowadays, we write this as the Hopf exact sequence

$$\pi_2(X) \to H_2(X) \to H_2(\pi_1(X)) \to 0.$$

(Of course the first homology of a space depends only on its fundamental group; it is the abelianization.)

We shall avoid any discussion of homological algebra and work purely topologically. Obstruction theory very quickly leads to the following fact:

Theorem   *For any countable group π, there is a space $B\pi$ whose fundamental group is p and whose universal cover is contractible. This space is unique up to pointed homotopy type. Moreover, for any homomorphism $\pi \to \pi'$ between groups, there is a unique pointed homotopy class of maps between spaces $B\pi \to B\pi'$ whose induced map on fundamental groups is the given homomorphism.*

This theorem embeds group theory within homotopy theory. As a result, any homotopy functor gives a functor on groups: so we can have homology and

---

[36] Essentially, the difference between a deterministic and nondeterministic Turing machine is not one of calculability, but one of speed; it measures the difference between discovering and verifying membership. Note that, in a group with an unsolvable word problem, not only is it hard to discover that a word is trivial, but it is also hard to verify that it is—the "certificate of triviality," for example, the product of relations one must use, can be noncomputably long.

cohomology, and more exotic things like K-theory and stable homotopy theory, whatever is useful.

Thus, we will write $H_i(\pi)$ for $H_i(B\pi)$ (and similarly for cohomology and coefficients, etc.). Some of the low groups have special interpretations.

Proposition *If A is abelian, then $H^1(\pi, A) = Hom(\pi, A)$. Also $H^2(\pi, A)$ is in a $1 : 1$ correspondence with central extensions of π by A, i.e., exact sequences*

$$0 \to A \to E \to \pi \to e.$$

(We denote the trivial group by e and the trivial abelian group by 0.)

The condition that the extension be central means that A lies in the center of E. Thus, for instance, $\mathbb{Z}_2$ has two central extensions by $\mathbb{Z}$: $\mathbb{Z} \times \mathbb{Z}_2$ and $\mathbb{Z}$ (mapping onto $\mathbb{Z}_2$); the infinite dihedral group $\mathbb{Z}_2 * \mathbb{Z}_2$ mapping onto $\mathbb{Z}_2$ is a noncentral extension by $\mathbb{Z}$.

If π is *perfect*, that is, $H_1(\pi) = 0$, meaning that π is its own commutator subgroup, then π has a universal central extension, that is, one that all others map through. For this extension, the center is $H_2(\pi)$, and it corresponds to the tautologous element of $H^2(\pi; H_2(\pi))$ that defines the Kronecker pairing from cohomology to the dual of homology.

Unraveling this a bit, under the assumption of perfection, we note that the universal coefficient theorem identifies $H^2(\pi; H_2(\pi)) = Hom(H_2(\pi); H_2(\pi))$; our element is the one that corresponds to the identity. Let us give a few simple examples of group homology.

Example 1 *The trivial group. Here $B\pi$ is a point, and homology vanishes above dimension zero.*

Example 2 *Free groups. If π is F_k, then $B\pi$ is a wedge of k circles, with $H_i = 0$ for $i > 1$, $H_0 = \mathbb{Z}$, and $H_1 = \mathbb{Z}^k$. It is worth mentioning in this example that when $B\pi$ is a finite complex, it makes sense to discuss its Euler characteristic, which is a well-defined integer (as the Euler characteristic is a homotopy invariant of finite complexes), and computable from the homology with coefficients in any field.*

Example 3 *Free abelian groups. If $\pi = \mathbb{Z}^k$, then $B\pi$ is T^k, the k-torus. Now, the Künneth formula can be applied to show that the homology of $B\pi$ is torsion-free of rank the binomial coefficient $k!/a!(k-a)!$.*

Example 4 *Cyclic groups. For $\pi = \mathbb{Z}_k$, it is a little harder to see what $B\pi$ looks like. One approach is to consider S^{2n-1} as the unit sphere of $\mathbb{C}^n$ on which $\mathbb{Z}_k$ acts freely, just by multiplying each coordinate by a primitive root of unity. As n gets large these quotient spaces resemble $B\pi$ more and more closely, and we can use their (co)homology to compute the group (co)homology.*

In this case, it is not hard to find a cell complex for the quotients $S^{2n-1}/\mathbb{Z}_k$ by induction on n. For $n = 1$, the quotient is a circle. For larger n, one gets a cell decomposition with one cell in every dimension until $2n - 1$. Explicit calculation then leads to

$$H_a(\mathbb{Z}_k) = \mathbb{Z}_k \quad H^a(\mathbb{Z}_k) = 0 \quad \text{with } a \text{ odd,}$$
$$H_a(\mathbb{Z}_k) = 0 \quad H^a(\mathbb{Z}_k) = \mathbb{Z}_k \quad \text{with } a \text{ even} > 0.$$

The Künneth formula can then be applied to give the calculation for abelian groups.

Remark. In general, one can build a $B\pi$ using a construction of Milnor. Consider the infinite join $\pi * \pi * \cdots$. It is contractible and has a free π action. The quotient is our desired space. (Recall that $X * Y$ is the space made up of lines connecting a point of X to a point of Y. The join of k discrete spaces is a wedge of $2k + 1$ spheres, up to homotopy type.) The chain complex obtained in this way for the computation of group homology is called the "bar resolution." (The n-chains are the free abelian group of n-tuples of group elements whose product is the identity.)

Before proceeding to methods of calculation, it seems worth mentioning a couple of applications. Throughout the rest of the book we will be giving many more geometric, but more involved, examples.

The first is to groups that act freely and properly discontinuously on Euclidean space. If π is such a group, then so is any subgroup. Note that if π so acts, it has a $B\pi$ which is finite dimensional, so its homology vanishes in all sufficiently large dimensions. In particular, by example 4, π cannot have any nontrivial finite cyclic subgroups, that is, π is torsion-free.

In fact, similar reasoning shows that if π acts freely on $\mathbb{R}^n$ or even a contractible manifold, the cohomology with arbitrary coefficients vanishes in dimension greater than n. If the quotient is noncompact, then even in dimension n one would get vanishing. If the quotient were compact, then the integral cohomology would be $\mathbb{Z}$ in dimension n; in fact, one could see that $H^n(\pi; \mathbb{Z}\pi) = \mathbb{Z}$ and vanishes otherwise. This last condition turns out to imply that $B\pi$ (if a finite complex) satisfies Poincaré duality.

It is an important conjecture that the converse might hold, namely, that given such a π there exists (a unique) free cocompact action on some contractible X iff $H^n(\pi; \mathbb{Z}\pi) = \mathbb{Z}$ and vanishes otherwise. (In the appendix to section 2.3, we will see, following M. Davis, that X cannot always be taken to be Euclidean space.)

Here is another rather different application. Suppose we are interested in the existence of short exact sequences of the form

$$1 \to F_k \to E \to \mathbb{Z}_r \to 0,$$

where E is torsion-free. Note that the kernel of the composite surjection $F_s \to \mathbb{Z} \to \mathbb{Z}_r$ is isomorphic to $F_{r(s-1)+1}$. So a sufficient condition is that $r \mid (k - 1)$.

Using homological ideas, one can show that the converse holds. The idea is that if E existed, we could consider BE. Let us suppose first that this is a finite complex. Then by covering space theory its r-fold cover would be homotopy equivalent to BF_k, a wedge of k circles. Since, by definition, the Euler characteristic is multiplicative in coverings, one would obtain that $1 - k = \chi(BF_k) = r\chi(BE)$, giving necessity. An actual proof goes like this. One shows that under these conditions, the chain complex of BE is still (chain equivalent to) a finite projective (over $\mathbb{Z}E$) chain complex. This allows one to use the Lefschetz fixed-point theorem for the $\mathbb{Z}_r$ action on BF. Then examination of the possible rational representations on H_1 forces the latter module to be a sum of one trivial summand and a number of copies of $\mathbb{Q}[e^{2\pi i/r}]$. This implies that $k \equiv 1 \bmod r$.

For our purposes, probably the most important calculational tool is the Mayer-Vietoris sequence of the amalgamated free product and HNN extension.

Theorem *If B is a subgroup of A and C (i.e., we are in the injective cases of the amalgamated free product and HNN extension) then*

$$B(A {\textstyle *\atop B} C) = BA \cup_{BB} BC,$$
$$B(A {\textstyle *\atop B}) = BA \cup_{BB \times \{0,1\}} BB \times [0, 1].$$

Consequently, one obtains exact sequences

$$\cdots \to H_k(B) \to H_k(A) \times H_k(C) \to H_k(A *_B C) \to H_{k-1}(B) \to \cdots,$$

$$\cdots H_k(B) \to H_k(A) \to H_k(A {\textstyle *\atop B}) \to H_{k-1}(B) \to \cdots,$$

where the map $H_k(B) \to H_k(A)$ in the second exact sequence is the difference of the two inclusions.

The proof of the theorem goes as follows. First, the right-hand sides have the right fundamental groups by van Kampen's theorem. One just wants to analyze the universal covers and see that they are contractible. This follows from the tree picture. The universal covers are made of copies of the universal covers of the BA's and BC's (which are themselves contractible) glued along universal covers of the BB's which are contractible. This guarantees the contractibility of the universal covers of $BA \cup_{BB} BC$ and of $BA \cup_{BB \times [0,1]} BB \times [0, 1]$.

Example 5 *Consider the group $G = \langle a, b, c, d \mid [a, b] = [c, d] \rangle$. It is clearly of the form $F_2 *_{\mathbb{Z}} F_2$ where the free groups are generated by $\langle a, b \rangle$ and $\langle c, d \rangle$. The relation is the amalgamation of a $\mathbb{Z}$ which is generated by the commutators. In this case the maps from the subgroup into the groups are trivial, so one gets the calculation that*

$$H_1(G) = \mathbb{Z}^4 \ \text{ and } \ H_2(G) = \mathbb{Z}.$$

The perspicacious reader probably noticed that this group is just the fundamental group of a surface of genus two, and we have computed the group *homology just by noticing that the surface is a BG! Therefore, it is worth noting that we would get the exact same calculation for group homology if we used $[a, b]^2 = [c^2, d^3]$. Indeed, for any words u, v in the commutator subgroups of $\langle a, b \rangle$ and $\langle c, d \rangle$, respectively, one would obtain for $\langle a, b, c, d \mid u = v \rangle$ the same homology. (However, these groups are not Poincaré duality groups, because they do not satisfy duality with respect to arbitrary coefficient modules.)*

A straightforward argument shows the following, which opens the way to applying spectral sequence techniques:

Proposition   *If one has a short exact sequence of groups $1 \to K \to E \to L \to 1$, then there is a fibration $BK \to BE \to BL$.*

As a special case (where the fibration is a circle bundle, and the spectral sequence becomes the Gysin sequence), one has for a central $\mathbb{Z}$-extension $1 \to \mathbb{Z} \to E \to L \to 1$ the sequence

$$\cdots \to H^{k-2}(L) \to H^k(L) \to H^k(E) \to H^{k-1}(L) \to \cdots .$$

We have chosen to write this sequence in cohomology because there one can interpret the map $H^{k-2}(L) \to H^k(L)$ concretely as cup product with the Euler class of the circle bundle.

Example 6   *Let us compute the homology of the Heisenberg group H of 3×3 upper triangular unipotent matrices. We have an exact sequence extension $1 \to \mathbb{Z} \to H \to \mathbb{Z}^2 \to 1$. The Euler class is the generator of $H^2(\mathbb{Z}^2)$. One thus obtains (via the Gysin sequence) that*

$$H_1(H) = \mathbb{Z}^2, \quad H_2(H) = \mathbb{Z}^2, \quad and \; H_3(H) = \mathbb{Z}.$$

Exercise

Write H as an HNN extension with $A = B = \mathbb{Z}^2$ and use a Mayer-Vietoris sequence to do the same calculation.

1.5 ARITHMETIC GROUPS

Arithmetic groups are groups that are defined similarly to $SL_n(\mathbb{Z})$, the group of invertible matrices with determinant one. They arise naturally all over mathematics, and they have been studied from many points of view.

In this section, we will review a few special theorems regarding the homology of arithmetic groups that we will need in chapter 4.

Consider a subgroup G of GL_n defined by polynomial relations with coefficients in the rational numbers $\mathbb{Q}$. In other words, we shall assume that there is a set of polynomials in the entries of the matrices and $\det^{-1}$ that define the group

G. It makes sense to discuss the $\mathbb{F}$ points of G, denoted $G(\mathbb{F})$, for any field $\mathbb{F}$ of characteristic 0. By $G_\mathbb{Z}$, we mean $G(\mathbb{Q}) \cap GL_n(\mathbb{Z})$. A discrete subgroup Γ of $G(\mathbb{Q})$ is called *arithmetic* if it is commensurable with $G_\mathbb{Z}$. Such subgroups are often *lattices*, that is, the natural metric on G/Γ has finite volume (see the theorem of Borel and Harish-Chandra below). For various reasons, it usually makes more sense to look at $K \backslash G / \Gamma$; for instance, it is an Eilenberg-MacLane space when Γ is torsion-free.

Note that given the real Lie group $G(\mathbb{R})$, there are many "$\mathbb{Q}$-forms," and these will give rise to different commensurability classes of arithmetic groups. For instance, we can define $O(n, 1)$ using any quadratic form over the rationals that has signature $(n, 1)$, and they will all give rise to arithmetic groups, but, unless these quadratic forms are homothetic over $\mathbb{Q}$, it is quite unlikely that these lattices will be commensurable. (A little below we will give a somewhat more general way of generating arithmetic lattices.)

We have already met a number of arithmetic groups: all finite groups, finitely generated abelian groups, finitely generated torsion free nilpotent groups (theorem of Malcev) such as the Heisenberg group, free groups (lie in $SL_2(\mathbb{Z})$), surface groups. Given a quadratic form f in n variables over the rationals, then $SO(n, f)$ defines a most interesting arithmetic group.

Remark. There is nothing sacred about $\mathbb{Q}$ in these definitions; using E, a finite extension of $\mathbb{Q}$, in its stead can be useful in defining more examples; in theory, this does not change the class of arithmetic groups, because if E is degree d over $\mathbb{Q}$, one can view $GL_n(E)$ as a subgroup of $GL_{nd}(\mathbb{Q})$. However, it is quite a bit simpler (and provides more insight) to write formulas using the general E rather than forcing them to be subgroups of $GL_{nd}(\mathbb{Q})$.

Here is an important example that, among other things, shows the need for a slight modification of the definition of arithmetic. Let us consider an orthogonal group of the quadratic form

$$x_1^2 + x_2^2 + \cdots + x_n^2 - \sqrt{2}x_{n+1}^2$$

for $\mathbb{Q}[\sqrt{2}]$. There are two embeddings of $\mathbb{Q}[\sqrt{2}]$ in $\mathbb{R}$. Thus $O(x_1^2 + x_2^2 + \cdots + x_n^2 - \sqrt{2}x_{n+1}^2, \mathbb{Z}[\sqrt{2}])$ is a lattice in $O(n, 1) \times O(n + 1)$(in the usual positive embedding, where $\sqrt{2}$ is positive, this quadratic form is of type $(n, 1)$; in the embedding where $\sqrt{2}$ is negative, the form is positive definite).

The image of this lattice in $O(n, 1)$ is a lattice there as well, because all that we are doing is projecting $O(n, 1) \times O(n+1) \to O(n, 1)$, which has a compact kernel. (The discreteness of the lattice means that we kill at most a finite normal subgroup of it when projecting.) The theorems we will explain presently show that the lattices just described are cocompact hyperbolic lattices. Torsion-free subgroups of finite index provide compact hyperbolic manifolds.

It is not at all obvious, but it is true, that arithmetic groups are finitely presented; they have a solvable word problem, are virtually torsion-free, and residually finite (this is a general fact about linear groups called Selberg's lemma).

Even the conjugacy problem is solvable in these groups (although not in all residually finite groups). However, as we saw above, the generalized word problem is usually not solvable in these groups.

Another remarkable property of these groups is that they are (Bieri-Eckmann) duality groups. This means that $H^k(\Gamma; \mathbb{Z}\Gamma)$ is nonzero for only one value of k. (This then implies a twisted Poincaré duality for Γ, where the twist is by that module.) This follows from the theory of Borel and Serre, which shows that the open manifold $K \backslash G / \Gamma$ can be compactified to a manifold with boundary, and their analysis of what the universal cover of its boundary looks like.

Recall that the *radical* of an algebraic group is its maximal connected algebraic solvable normal subgroup. The *(E-)rank* of $G(E)$ is the dimension of the maximal split torus (i.e., products of GL_1) defined over E that can be embedded in $G(E)$. G is *semisimple* if its radical is trivial.

Theorem (Borel and Harish-Chandra) *Let Γ be an arithmetic subgroup of G.*

1. *$G(\mathbb{R}) / \Gamma$ has finite G-invariant volume iff there are no $\mathbb{Q}$-homomorphisms from the identity component of G to GL_1.*
2. *$G(\mathbb{R}) / \Gamma$ is compact iff G has no subgroup isomorphic to GL_1 which is iff it has finite volume and every unipotent element of $G(\mathbb{Q})$ lies in its radical.*

For example, $GL_m(\mathbb{Z})$ is not a lattice in $GL_n(\mathbb{R})$ because of the homomorphism to $\mathbb{R}^*$ (= GL_1) given by the determinant. Recall that unipotents are elements differing from the identity by a nilpotent. Condition 2 is equivalent to saying that the $\mathbb{Q}$-rank of G is 0.

We can use this theorem to check that the hyperbolic lattices produced above are actually cocompact. Since we are using embeddings of $\mathbb{Q}[\sqrt{2}]$ in $\mathbb{R}$, an element of the lattice is unipotent iff it is under either embedding. However, the embedding where $\sqrt{2}$ is negative gave rise to the orthogonal group, which has no nontrivial unipotents; a fortiori, neither does the lattice.

Example *For quadratic forms, (1) holds unless $n = 2$ and f represents 0 (i.e., there are nontrivial nullvectors in E), and (2) holds whenever the form is anisotropic (i.e., has no nullvectors).*

We shall confine the rest of our discussion to the semisimple case, and, as indicated above, we shall extend the definition of an arithmetic subgroup of a real Lie group H to be the image of an arithmetic group in a group G defined over $\mathbb{Q}$ under a Lie homomorphism from G onto an open subgroup of H which has a compact kernel.

Every semisimple group has a $\mathbb{Q}$-form that gives it arithmetic lattices. In fact, G contains both uniform (= cocompact) and nonuniform lattices.

Theorem *(Margulis's arithmeticity.) If $\mathbb{R}$-rank$(H) > 1$, then all irreducible lattices in H are arithmetic.*

In rank one, the existence of nonarithmetic lattices depends strongly on the Lie group and has been the object of intensive study. For instance, all the $SO(n, 1)$'s have nonarithmetic lattices, but $Sp(n, 1)$ does not. It is unknown whether $U(n, 1)$ has such lattices when n is large.

Later, I will be interested in the cohomology of certain arithmetic lattices. While I cannot go into the details here, it seems worth just *mentioning* some of the ideas that have been brought to bear on this problem. Everything is a lot simpler in the uniform (i.e., cocompact) case, although, with more work, similar results can often be obtained for the nonuniform case.

The first key point is that for any semisimple group the coset space $K\backslash G$ with its right invariant metric has nonpositive curvature.[37] Consequently, it is a Euclidean space, and the manifold[38] $K\backslash G/\Gamma$ is a $B\Gamma$. Thus, the group homology is the study of the homology of this manifold. The following discussion is more straightforward if we assume that Γ is a uniform lattice, that is, that $K\backslash G/\Gamma$ is compact.

For convenience, we switch to cohomology and make use of an essentially tautological isomorphism:

$$H^*(\Gamma; \mathbb{C}) = H^*(K\backslash G/\Gamma; \mathbb{C}) = H^*(\mathfrak{g}, \mathfrak{k}; C^\infty(G/\Gamma))$$

where the last term is Lie algebra cohomology; the isomorphism is a consequence of a cochain complex isomorphism between the deRham model of the cohomology of $K\backslash G/\Gamma$ and the defining complex of relative Lie algebra cohomology with coefficients.

Now, one can show that the decomposition of $L^2(G/\Gamma)$ as a sum of irreducible representations,

$$L^2(G/\Gamma) \approx \bigoplus m(\pi, \Gamma)H_\pi$$

(where the m's are multiplicities and the H_π are the irreducible Hilbert space representations of G that are summands of the regular representation of G), gives one of $H^*(\mathfrak{g}, \mathfrak{k}; C^\infty(G/\Gamma))$ as well. That is,

$$H^*(\Gamma; \mathbb{C}) \approx \bigoplus m(\pi, \Gamma)H^*(\mathfrak{g}, \mathfrak{k}; H_\pi). \qquad (2)$$

(This is essentially some kind of smoothing theorem, analogous to the fact that smooth singular homology and the continuous version are isomorphic.)

This result is the *Matsushima formula*. It gives a lot of useful information, including useful vanishing and nonvanishing theorems. A very useful result is that for lattices in semisimple groups, through some range linear in the $\mathbb{R}$-rank, the terms not coming from the trivial representation give vanishing contributions. This means that the cohomology (in some range) is independent of the lattice! (The cohomology associated with the trivial representation is isomorphic to that of the *compact dual* of $K\backslash G$.)

---

[37] We will discuss the elementary geometry necessary to follow this discussion in chapter 3. In any case, the trusting reader can just skip a sentence or so.
[38] Orbifold, if Γ has torsion.

I should emphasize that this is true only for coefficients with characteristic 0. For finite coefficients, one can see that the opposite is almost always true. (Hint: Think about lattices corresponding to p-Sylow subgroups of finite quotients of a given lattice.) It also fails strongly around the rank, as one can see for surfaces.

Remarks

Although our discussion assumed cocompactness, there are versions of the Matsushima formula and the vanishing theorems that are true for general lattices. Moreover, one can also generalize a great deal of the theory of arithmetic groups (including hard things like arithmeticity and cohomology calculations, although not the soft parts like finite generation!) to "S-arithmetic groups." These are lattices in products of real and p-adic fields, that is, groups like $\mathrm{SL}_n(\mathbb{Z}[1/k])$ for an integer k.

One of the most striking results of this development is that for any number field E (finite extension of $\mathbb{Q}$)

$$H^*(G(E)) = H^*_{\mathrm{cont}}(G_\infty) \tag{3}$$

where the subscript "cont" means continuous cohomology and G_∞ refers to the copies of G at the infinite places. In other words, it looks as if equation (2) holds, but with no contributions of any of the other representations besides the trivial one!

The formula (3) is based on the fact that, by considering all of the E-points, one has essentially arranged for the rank of the "lattice" to be infinite.

A second useful method is L^2-cohomology. (The applications of this considerably transcend the study of lattices.) While the results are not quite precise, they give conclusions such as that, if $K \backslash G$ is odd dimensional, the Betti numbers of regular covers of a given lattice grow as $o(\text{volume})$ (i.e., sublinearly in the index of the cover), and in even dimensions, all but the middle cohomology groups do the same. On the other hand, the middle-dimensional groups do have ranks that are asymptotic to a multiple of the volume. (Some people even believe that this behavior is typical of residually finite groups that are fundamental groups of aspherical manifolds.) The drawback of this method is that it is hard to go from an L^2 calculation back to an ordinary calculation.

Later on, we will need sharp information about vanishing and nonvanishing of cohomology for negatively curved manifolds. This necessitates a look at lattices of $\mathbb{R}$-rank one. A very deep theorem of Clozel that gives sharp vanishing and nonvanishing results for a class of arithmetic lattices in $U(n, 1)$.

Theorem *For every n, there are complex hyperbolic n-manifolds[39] whose cohomology is nonzero in exactly the following dimensions:*

  1. there is a rank-one piece in every even dimension $0 \leq d \leq 2n$;

---

[39]These are of complex dimension n, and thus of real dimension $2n$.

2. *for any divisor a of n + 1 (less than n + 1) there are elements in every second dimension between n − a + 1 and n + a − 1.*

It is worth making a couple of comments about this theorem. First, the relative Lie algebra cohomology is nonzero in the vanishing range here; the theorem is an arithmetic phenomenon, and, indeed, it is known that it fails for other lattices in $U(n, 1)$. (Of course, what is happening is that the multiplicities occurring in the Matsushima formula are zero.)

Second, there are a large number of contributions to the proof of this theorem coming from deep number theory, à la Langlands' program. While I cannot say anything that really elucidates what is going on, I should probably mention that the "baby example" of these ideas is Deligne's proof of the Weil conjectures. These conjectures give an arithmetic method for computing the cohomology of smooth projective varieties. According to this work, the number of points on the variety over the various finite fields contains exactly the same information as the rational cohomology. The cohomology of $\mathbb{CP}^n$ "corresponds" to the number of points in $\mathbb{P}^n(\mathbb{F}_q)$ being $1 + q + q^2 + \cdots + q^n$.

In fact, $\mathbb{CP}^n$ is the compact dual of complex hyperbolic n-space, $U(n + 1, 1)/U(n) \times U(1)$. The classes accounted for in (1) are the classes coming from the trivial representation in the Matsushima formula, that is, the classes from the compact dual. Geometrically, the dual homology classes can be thought of as intersections of the complex hyperbolic manifold, of a smooth projective variety, with linear subspaces of $\mathbb{CP}^n$. The other classes are harder to account for, although their general placement symmetrically around the middle is Poincaré duality, their upward growth toward the middle is the Lefschetz theorem, and the nonvanishing in the middle can be seen, in even complex dimension, using the Hirzebruch signature theorem, and by the L^2-method.

1.6 REALIZATION OF SEQUENCES OF GROUPS AS GROUP HOMOLOGY

While we will not need the full depths of the following theorems, they are very interesting, and the special cases that we will need are not substantively simpler than the general case.

The basic issue we are interested in is the appearance of the sequence of homology groups of a finitely presented group. Given the Higman embedding theorem, it is perhaps not surprising that there is a strong logical component to this problem. On the other hand, the reader might be surprised to find that, for instance, there is a group G such that

1. for each a, $H_a(G) = 0$ or $\mathbb{Z}$, and
2. $\{ a \mid H_a(G) = 0 \}$ is neither c.e. nor the complement of a c.e. set.

We will also see that there is a finitely presented group G such that $H_a(G) = \mathbb{Z}_a$ or $H_a(G)$ is a sum of copies of $\mathbb{Z}$, where the number of copies is the ath digit of $\pi + e^2$.

In fact, there is an almost complete solution to this problem, but as of yet there does not seem to be one to the natural question of what cohomology algebra structures can exist. One has to be careful about the exact formulation of this question, because it is not yet known even for spaces. In fact, the question should be formulated in a way that explicitly compares what happens for groups to what happens for spaces.

There is a sense in which groups are no more special than general spaces:

Theorem   *For any connected simplicial complex X, there is a group π and a map $f : B\pi \to X$, which is an isomorphism on homology. In fact, for any covering space of X, the map from the induced cover of $B\pi$ is also an isomorphism on homology.*[40] *Moreover, if X is a finite complex, $B\pi$ can also be taken to be a finite complex. If X is a countable complex, π can be taken countable. Moreover, if X is a c.e. space, then π is a c.e. group.*

We will return to the precise meaning of the c.e. group and c.e. space. Let us concentrate on the proof of the other parts of the theorem.

The construction of $B\pi \to X$ has two steps. The first is the construction of "n-simplices of groups." The second is merely the assembly of these according to the same data that one uses to assemble standard simplices to construct X.

To begin, one needs a nontrivial acyclic group A (that is, a group whose reduced homology vanishes in all dimensions). One can do so using an injection $F_4 \to F_2$ that looks like the projection on homology, and then producing an amalgamated free product $F_2 *_{F_4} F_2$, where the two injections of $F_4 \to F_2$ are such maps, just arranged to be projections to different factors.

Using A, we can easily build a 1-simplex of acyclic groups, using A for each of the two vertices and $A \times A$ for the group associated with the 1-simplex. (Note that here we map the group associated with a vertex into the group associated with an edge, the opposite of what we did with the graph of a group.) Using amalgamated free products and HNN extensions, one can assemble these groups to build a π, such that $B\pi \to G$ for any connected graph G, and by the Mayer-Vietoris exact sequences in section 1.4, this map is an isomorphism in homology (and, by the exact same argument, the same holds for covers). This construction proves the theorem for graphs.

Now, to do two-dimensional complexes, one needs to construct a 2-simplex. That is, we need an acyclic group B that contains the result of applying the construction to a circle, thought of as a triangle, that is, as the boundary of a 2-simplex (and similarly in higher dimensions.)

---

[40]This notion can be most succinctly described in terms of the "plus construction," which will be explained in chapter 2.

We shall not give a construction of these simplices (all such contructions that I know about are somwhat tricky), and shall instead, rely on the paper [BDH] mentioned in the notes section. In any case, I hope the idea is clear.

Now, let us move on to the notions of a c.e. group and c.e. space. A c.e. group is a group with generators $x_1, x_2, x_3, \ldots$ and a set of relations that is a c.e. set. In other words, there is a Turing machine that constructs the relations. Note that it is entirely equivalent to ask that the set of relations that hold be c.e. or to give a c.e. generating set for these relations. To define the notion of a c.e. space, we will think of the vertices as being the integers (or a finite set of them). We can think of the simplices as being $(n + 1)$tuples of vertices, which can be encoded by natural numbers. So a simplicial complex is just some set of tuples of natural numbers (with the additional property of being closed under inclusion). We shall suppose that our complexes are effectively connected.

Note that the homology groups of a c.e. space are actually c.e. abelian groups. (Hint: First check that a c.e. abelian group, up to computable isomorphism, is equivalent to a c.e. sequence of finitely presented abelian groups with (c.e.) homomorphisms from one group to the next.)

We leave it to the reader to check that the above constructions produce c.e. groups from c.e. spaces.

The space $B\pi$ for π finitely presented (or even c.e.) groups is actually a c.e. simplicial complex, as one can see by going carefully through the Milnor construction. This suggests the following:

Conjecture *For X an effective simplicial complex, whose 2-skeleton is finite (up to homotopy), there is a finitely presented π, and a map $B\pi \to X$, which is an isomorphism on homology.*

This would lead to a characterization of the sequences of groups that can be the homology groups of a finitely presented group. They would be the c.e. sequences of c.e. abelian groups whose first two groups are finitely generated. The following theorem implies something that is quite close.

Theorem *If X is any c.e. simplicial complex, then there is a finitely presented π, and a map $f : B\pi \to \Sigma^2 X$ to the second suspension of X, which is an isomorphism on homology.*

First we find a c.e. group π that resembles X. Then we can embed π in a universal acyclic group U,[41] and form $\pi' = U *_\pi U$, which resembles the suspension of X. This is now a finitely generated c.e. group, which we will denote π'. π' also embeds in U. $U *_{\pi'} U$ resembles the second suspension, and is also finitely presented, proving the theorem.

Corollary *For $n > 3$, for any c.e. abelian group A, there is a f.p. group π such that $H_a(\pi) = A$ for $a = n$, and is 0 otherwise.*

---

[41]Recall that a universal group is a finitely presented group that contains all others. Baumslag, Dyer, and Miller constructed acyclic universal groups.

Proof. A is a limit of a c.e. sequence of finitely presented abelian groups. Thus, one can form the c.e. sequence of Moore spaces and maps to produce a c.e. space $M(A, n - 2)$. Setting $X =$ the limit of these spaces and applying the theorem gives the result.

Remark. To put some more flesh onto the above proof, we should make a few simple remarks. Recall that a Moore space of type (A, k) is a simply connected space all of whose homology groups vanish except for H_k, and $H_k = A$. They exist for $k > 1$, and are unique up to homotopy equivalence. For any homomorphism between A and B, there is a map from $M(A, k) \to M(B, k)$ inducing this homomorphism. (The homotopy class of maps is not unique, however.)

Note also that one can interpolate between two triangulations of a polyhedron P by a triangulation of $P \times [0, 1]$. Arbitrary homotopy classes can also be realized by simplicial maps, which allows one to build a c.e. space from the c.e. sequence of homotopy types and (the not quite well defined) sequence of homotopy classes of maps.

Corollary *If A_i is any c.e. sequence of c.e. abelian groups, such that A_1 and A_2 are finitely generated and A_3 is "untangled" in the sense of [BDM] (see the notes), then there is a finitely presented group with the A's as its homology sequence.*

An abelian group is "untangled" if it has a presentation with a c.e. basis for its relations. This condition is not necessary, and thwarts a complete characterization of the homology sequence of f.p. groups.

Proof. To realize the $A_k, k > 3$, one simply uses as X the wedge of the Moore spaces discussed in the previous proof. We can then take the free product with a group realizing the first three groups from [BDM] (and 0 above dimension three), to obtain our desired π.

NOTES

The elementary topology of fundamental groups, covering spaces, and van Kampen's theorem is all very nicely explained in

W. Massey. *Algebraic Topology: An Introduction.* Reprint of the 1967 edition. Graduate Texts in Mathematics 56. Springer-Verlag, New York, 1977.

A good introduction to basic combinatorial group theory and to the theory of group actions on trees can be found in

P. Scott and C.T.C. Wall. *Topological methods in group theory. In Homological Group Theory.* Proceedings of the Symposium (Durham, 1977), 137–203. London Mathematical Society Lecture Note Series 36. Cambridge University Press, Cambridge, 1979.

J. P. Serre. *Trees*. Translated from the French by John Stillwell. Springer-Verlag, Berlin, 1980.

A very nice survey article detailing a proof of the Novikov-Boone theorem (based on ideas of Cohen and Anderaa) is

J. Stillwell. *The word problem and the isomorphism problem for groups*. Bull. Amer. Math. Soc. 6 (1982), 33–56.

Textbook references can be found in Manin's book on logic referred to in the references to the last chapter (which relies on the solution to Hilbert's tenth problem on Diophantine equations) and Rotman's book, which gives a version of the methods of Boone and Novikov.

J. Rotman. *An Introduction to the Theory of Groups*, 4th ed. Graduate Texts in Mathematics 148. Springer-Verlag, New York, 1995.

The Adian-Rabin theorem can be found in

M. Rabin. *Recursive unsolvability of group theoretic problems*. Ann. Math. (2) 67 (1958), 172–194.

C. Miller has written excellent surveys of all the above work on decision problems, and the material in the section on "refinements and extensions."

C. Miller III. *On Group-Theoretic Decision Problems and Their Classification*. Annals of Mathematics Studies 68. Princeton University Press, Princeton, N.J. University of Tokyo Press, Tokyo, 1971.

_____. *Decision problems for groups: Survey and reflections*. Algorithms and Classification in Combinatorial Group Theory (Berkeley, Calif., 1989), 1–59. Math. Sci. Res. Inst. Publ. 23. Springer-Verlag, New York, 1992.

The result we mentioned about a solvable group with an unsolvable word problem is due to O. Kharlampovich. It can be found in the survey paper of Miller.

The idea of a Dehn function is an old one, and in recent years it has been studied with renewed vigor. The papers that most directly influenced our discussion in section 1.3 are

S. Gersten. *Dehn functions and ℓ^1-norms for finite presentations*. Algorithms and Classification in Combinatorial Group Theory (Berkeley, Calif., 1989), 195–224. Math. Sci. Res. Inst. Publ. 23, Springer-Verlag, New York, 1992.

M. Gromov. *Asymptotic invariants of infinite groups*. Geometric Group Theory, vol. 2. Edited by G. Niblo and M. Roller. London Mathathematical Society Lecture Note Series 182, Cambridge University Press, Cambridge, 1993

and the beautiful survey paper

A. Olshanski and M. Sapir. *Length and area functions on groups and quasi-isometric Higman embeddings*. Preprint, arXiv: math.GR/9811107,

which explains the intricate construction of Sapir with Birget and Rips on Dehn functions, as well as related work on length functions, and distortion functions of group embeddings.

As we move on to group homology, this is a standard technique. A good textbook reference is

K. Brown. *Cohomology of Groups*, 2nd ed. Graduate Texts in Mathematics 87. Springer-Verlag, New York, 1994.

Other general references for homological algebra are

S. MacLane. *Homology*. Springer-Verlag, New York, 1963.

C. Weibel. *An Introduction to Homological Algebra*. Cambridge University Press, Cambridge, 1994.

An excellent survey article on the topic of arithmetic groups, and a good place to start, is

J.P. Serre. *Arithmetic groups*. In *Homological Group Theory*. Proceedings of the Symposium. (Durham, 1977), 105–136. London Mathematical Society Lecture Note Series 36. Cambridge University Press, Cambridge, 1979.

More of a sense of the subject can be gleaned from the following:

C. L. Siegel. *Lectures on the Geometry of Numbers*, notes by B. Friedman. Rewritten by Komaravolu Chandrasekharan with the assistance of Rudolf Suter, with a preface by Chandrasekharan. Springer-Verlag, Berlin, 1989.

A. Borel. *Introduction aux Groupes Arithmétiques*. (French) Publications de l'Institut de Mathématique de l'Université de Strasbourg, XV. Actualités Scientifiques et Industrielles 1341. Hermann, Paris, 1969.

_____. *Algebraic Groups and Discontinuous Subgroups*. Proceedings of the Symposium in Pure Mathematics (Boulder, Colo., 1965). American Mathematical Society, Providence, R.I., 1966.

G. Margulis. *Discrete Subgroups of Semisimple Lie Groups*. Ergebnisse der Mathematik und ihrer Grenzgebiete 17. Springer-Verlag, Berlin, 1991.

V. Platonov and A. Rapinchuk. *Algebraic Groups and Number Theory*. Pure and Applied Mathematics. 139. Academic Press, Boston, Mass., 1994.

R. Zimmer. *Ergodic Theory and Semisimple Groups*. Monographs in Mathematics 81. Birkhäuser, Basel, 1984.

D. Witte. *Introduction to Arithmetic Groups* (in preparation, see http://www.math.okstate.edu/~dwitte).

These are very useful general sources (although the first couple are definitely dated). The finite presentation of arithmetic groups is due to Raghanuthan; the duality properties are due to Borel and Serre, and are based on the theory of Tits buildings, and the work of Solomon and Tits. Their paper is

A. Borel and J. P. Serre. *Corners and arithmetic groups*, Comment. Math. Helv. 48 (1973), 436–491.

The solution to the conjugacy problem for arithmetic groups is a special case of much more general algorithms that can be found in

F. Grunewald and D. Segal. *Some general algorithms: I. Arithmetic groups.* Ann. Math. (2) 112 (1980), no. 3, 531–583.

The arithmeticity theorem of Margulis is proven in the references by Margulis and Zimmer above.

The Matsushima formula and its systematic application to the calculation of cohomology of arithmetic groups is the basis of the book

A. Borel and N. Wallach. *Continuous Cohomology, Discrete Subgroups, and Representations of Reductive Groups*, 2nd ed. American Mathematical Society, Providence, R.I., 2000,

which also describes the work of other authors on this topic. Formula (3) is the last formula of that book, and first appeared in a paper of Borel and J. Yang (which depends strongly on a paper of J. Franke).

L^2-cohomology of universal covers is a big business, which I cannot do justice to, even in summary. The results about lattices that I referred to are proven in

J. Cheeger and M. Gromov. *Bounds on the von Neumann dimension and L^2 cohomology and the Gauss-Bonnet theorem for open manifolds*, J. Diff. Geom. 21 (1985), 1–34.

It is a striking general fact, due to Lück, that for residually finite groups, the von Neumann dimensions of the L^2-cohomology of the universal cover is the same as the limit of the renormalized Betti numbers of a descending chain of normal finite index subgroups.

W. Lück. *Approximating L^2 invariants by their finite analogues.* Geom. Funct. Anal. 4 (1994) 455–481.

Lück has written an excellent survey book on the many topics that L^2-cohomology of covers sheds light on, and so atones for the brevity of my discussion here.

W. Lück. *L^2 invariants: Theory and Applications to Geometry and K-Theory.* Springer-Verlag, Berlin, 2002.

Clozel's paper is

L. Clozel. *On the cohomology of Kottwitz's arithmetic varieties.* Duke Math. J. 72 (1993), 757–795.

(See also his paper *Representations galoisiennes associées aux representations automorphes autoduales de GL(n)*, Publ. Math. d'IHES 73 (1991), 97–145, for the proof that $U(n + 1, 1)$ contains lattices of the sort that are analyzed in the 1993 paper.)

That groups have the same homology as arbitrary spaces is the Kan-Thurston theorem. The proof we sketched is from the paper

G. Baumslag, E. Dyer, and A. Heller. *The topology of discrete groups.* J. Pure Appl. Algebra, 16 (1980), 1–40,

which gave the first proof of the finiteness statement of the theorem. The fact that the construction of π is c.e. if the space X is a c.e. simplicial complex is implicit in that paper. Hausmann has gone further and shown that if X is finite the group π in this theorem can be taken to be a duality group in the sense of Bieri and Eckmann; see

J. C. Hausmann. *Every finite complex has the homology of a duality group.* Math. Ann. 275 (1986), no. 2, 327–336.

Most of the remaining results of this section are due to

G. Baumslag, E. Dyer, and C. Miller III. *On the integral homology of finitely generated groups.* Topology 22 (1983), 27–46.

The way these papers are combined to squeeze out a drop of extra information is taken from

A. Nabutovsky and S. Weinberger. *The fractal geometry of Riem/Diff I.* Geometriae Dedicata 101 (2003), 1–54.

Chapter Two

Designer Homology Spheres

This part begins our transition to geometry. We shall study some aspects of the theory of homology spheres because of their roles as "doppelgangers" for the sphere; they are therefore useful for producing doppelgangers for arbitrary manifolds. We begin with some classical and elementary theorems about homology spheres, then prove Novikov's[42] theorem on the algorithmic nonrecognizability of the sphere (and of any manifold) in high dimensions. Then we come to our first application of the logical method, Nabutovsky's thesis, which shows how hard it is to uncrumple a hypersphere in the sphere. (We shall also see stronger versions of this phenomenon for nonsimply connected hypersurfaces.)

Afterward we return to the general theory of homology spheres and their classification and use these to produce homology spheres with properties that are useful to us. In the appendixes, we review some surgery theory and facts about the Novikov conjecture that will be useful later.

2.1 FUNDAMENTAL GROUPS

We begin this story with Smale's celebrated theorem, which earned him the Fields medal, and which is at the basis of almost all of high-dimensional topology.

Theorem *A compact smooth manifold is diffeomorphic to D^n $(n > 5)$ iff it is contractible (i.e., iff it is simply connected and acyclic) and its boundary is simply connected. A compact manifold is an annulus $S^{n-1} \times [0, 1]$ iff it deform retracts to each of its boundary components, and one of them is a sphere.*

In fact, for M simply connected, one can characterize $M \times [0, 1]$ in the analogous way.[43] There are also versions of this theorem in the PL and topological categories. It implies

Corollary 1 *A closed n-manifold Σ^n, where $n > 4$, is PL-homeomorphic to S^n iff it is homotopy equivalent to S^n, that is, iff it is simply connected and has the same homology as S^n.*

---

[42]This Novikov is the topologist/physicist Novikov, son of the Novikov who showed the algorithmic undecidability of the word problem.

[43]This is called the h-cobordism theorem and will be discussed below.

Definition. A closed n-manifold Σ is a homology sphere if $H_a(\Sigma) = H_a(S^n)$ for all a.

The corollary can be phrased as asserting that a simply connected homology sphere of dimension at least five is PL-homeomorphic to the sphere. That a simply connected homology sphere is homotopy equivalent to the sphere can be seen directly from the Hurewicz isomorphism theorem and the Whitehead theorem.

The corollary follows from the theorem by removing a little ball from Σ, and identifying the complement as a ball using Smale's theorem.[44] The union of two balls might or might not be a sphere; it depends on how they are glued together. The issue is whether the diffeomorphism of S^{n-1} extends over D^n. However, by radial extension, any PL-homeomorphism extends over the ball, so there is no issue in the PL category.

As is well known, there are exotic differential structures on the sphere, so the corollary is not correct as stated in the smooth category. It turns out that the oriented diffeomorphism types of smooth structures on the sphere form an abelian group (under #), which can be effectively studied by surgery theory. (We will review surgery theory in an appendix to section 2.4.)

Smale's theorem has another corollary:

Corollary 2 *Any smooth embedding of S^{n-1} in S^n for $n > 4$ is, up to reparametrization, isotopic to the standard embedding of the equator in S^n.*

One easily sees that the complementary regions are disks, and the disk has a unique embedding up to isotopy (an isotopy to a linear embedding can be obtained by shrinking back to a small ball that one can then analyze by advanced calculus). In section 2.3, we shall see that the isotopy must necessarily be quite complicated looking.

Having dispensed, for the time being, with simply connected homology spheres, let us begin our study of nonsimply connected ones.

Theorem *For $n > 4$, a group π is the fundamental group of a homology n-sphere iff*

1. *it is finitely presented;*
2. *$H_1(\pi) = 0$; and*
3. *$H_2(\pi) = 0$.*

One can arrange for this homology sphere to bound a contractible manifold. As groups satisfying the condition 2 are called perfect, groups satisfying (2) and (3) are called superperfect.

The first two conditions are obvious. Condition 3 follows from the Hopf exact sequence (of chapter 1, section 2.1) $\pi_2(\Sigma) \to H_2(\Sigma) \to H_2(\pi_1(\Sigma)) \to 0$. Since $H_2(\Sigma) = 0$, then $H_2(\pi) = 0$.

---

[44]The argument works for $n > 5$. For $n = 5$ the necessary argument is trickier.

Remark. These conditions are necessary in lower dimensions > 1, but are not sufficient for $n = 3, 4$.

We shall describe the proof only for $n > 5$; the case $n = 5$ is a bit trickier. Let $\pi = \langle g_1, g_2, g_3, \ldots, g_k \mid r_1, \ldots, r_m \rangle$. Consider $V = S^1 \times S^{n-1} \# \cdots \# S^1 \times S^{n-1}$ a total of k times. Its fundamental group is free on k generators, which we shall call $g_1, g_2, \ldots, g_k$. Now, each word $r_1, \ldots, r_m$ represents an element of the fundamental group and thus is represented by a map of S^1 into V. We can smoothly approximate this map, and then apply the Whitney embedding theorem (or transversality) and make these disjoint smooth embeddings. The tubular neighborhood theorem provides each of these curves $\gamma_1, \gamma_2, \ldots, \gamma_k$ with neighborhoods diffeomorphic to $S^1 \times D^{n-1}$. Let $W = V - \cup \gamma_a \times D^{n-1} \cup D^2 \times S^{n-2}$. One readily computes that

1. $\pi_1(W) = \pi$;
2. $H_2(W) = \mathbb{Z}^{m-k}$; and
3. $H_m(W) = 0$ otherwise for $m < n - 2$.

(The first requires $n > 3$; the second and third use condition 2 and several Mayer-Vietoris sequences.) Now we will use $n > 4$. By the Hopf exact sequence, we have

$$\pi_2(W) \to H_2(W) \to H_2(\pi) = 0.$$

We can represent generators of $H_2(W) = \mathbb{Z}^{m-k}$ by spheres which, as before, we can assume to be embedded and to have neighborhoods diffeomorphic to $S^2 \times D^{n-2}$.

We do the same trick: remove neighborhoods of these spheres and glue in copies of $D^3 \times S^{n-3}$. This kills H_2 (note that here we use $n > 5$, because for $n = 5$ there are new "dual" copies of S^2 produced as the copies of S^{n-3}) and, because we were careful to use a basis, it does not produce any H_3. The result is our Σ.

If we started off with the boundary connected sum of copies of $S^1 \times D^n$ and then glued on copies of $D^1 \times D^{n+1-i}$ instead of removing and gluing in, we would produce Σ simultaneously with a contractible manifold that it bounds.

Remark. Note that the proof is essentially effective. Steps that could require analysis are (1) how thin the product neighborhoods around the various spheres produced are (this will affect curvature calculations and other geometric measurements), and (2) how hard it is to produce the copies of S^2 that come out of the Hopf exact sequence. In general, this step is computationally quite intensive, and we will discuss it again later. In any case, once one knows that something exists, as a theorem, its construction can always be done in computable time![45]

---

[45]There is an interesting subject that analyzes how fast functions constructed in this existential way are in terms of the strength of the axiom systems that produce them; an existence proof that uses "weak axioms" actually produces objects that can be constructed "fairly rapidly."

Remark. There is a similar theorem characterizing the fundamental groups of manifolds homology equivalent to a given manifold M^n with $n > 4$, but its proof is more involved. If $\Gamma \to \pi_1(M)$ is a surjection whose kernel is superperfect, then there is a homology equivalence $M' \to M$ (with coefficients in an arbitrary $\mathbb{Z}[\pi_1(M)]$-module) and with $\pi_1(M') = \Gamma$.

The following theorem gives a connection between the aggregate of all homology spheres and the simply connected ones.

Theorem *If Σ^n with $n > 3$ is a homology sphere, then one can assign to Σ^n a smooth structure on S^n which is trivial iff Σ^n is the boundary of a contractible manifold.*

By a smooth structure on S^n one means a smooth manifold θ_n that is PL-homeomorphic to S^n; the structure is trivial if this manifold is diffeomorphic to S^n. Another way to state the theorem is that there is a unique smooth structure on the sphere θ such that $\Sigma \# - \theta$ bounds a contractible manifold. Sometimes people describe this as saying, "it is possible to change the smooth structure of S in the neighborhood of a point to make it bound a contractible manifold."

Remark. This result is false in dimension three because of Rochlin's theorem which restricts the quadratic forms of smooth 4-manifolds; the Poincaré homology sphere (= dodacahedral manifold) is an example.

As usual, we will skip the trickier case of $n = 4$, and then the proof is a simple variant of the proof of the previous theorem. Consider $\Sigma \times [0, 1]$. Let $\gamma_1, \gamma_2, \ldots, \gamma_k$ be curves in $\Sigma \times \{1\}$ with neighborhoods diffeomorphic to $S^1 \times D^{n-1}$ which generate $\pi_1(\Sigma)$. Attach to $\Sigma \times \{1\}$ copies of $D^2 \times D^{n-1}$ along the copies of $S^1 \times D^{n-1}$. This produces a new simply connected manifold, with two boundary components, such that one is Σ and the other has $H_2 = \mathbb{Z}^k$. Now find 2-spheres which form a basis for this H_2, and "kill them" by attaching $D^3 \times D^{n-2}$ to them. One checks that this produces a new manifold K such that:

1. $K = S \cup \theta$, where θ is a homotopy (= simply connected homology) sphere.
2. K deform retracts to θ.

We leave it to the reader to use the annulus part of Smale's theorem to see that θ is well defined up to diffeomorphism. If ∂K is the sphere, we can glue a D^n onto K to obtain a contractible manifold.

Corollary *Every homology sphere S^n for $n > 4$ embeds in S^{n+1} after changing its smooth structure in the neighborhood of a point. There is an (actually unique) embedding (up to reparametrization and isotopy) which is the fixed set of a smooth involution S^{n+1}.*

Make Σ^n the boundary of a contractible manifold K (at the cost of a change in smooth structure). Then $K \times [0, 1]$ is readily seen to be a disk (by Smale's theorem). It has an involution which restricts on the boundary to one on S^{n+1} with Σ as fixed-point set.

2.2 ALGORITHMIC IMPOSSIBILITY RESULTS

Theorem *For no compact M^n, where $n > 4$, is there an algorithm to decide whether another manifold is diffeomorphic (or PL-homeomorphic or homotopy equivalent) to M.*

Remark. In the statement of this theorem, one can assume that the manifold is "given" in some combinatorial fashion. Here are some choices. One could be given a triangulation. One could be given a finite set of polynomials with rational (or integer) coefficients in several variables whose zero locus is the manifold. Finally, one could be given a handlebody decomposition. Morse theory, Nash's theorem on real algebraic structures, and the basic theory of semialgebraic sets interrelate these notions.

Proof. This is quite simple. Let A be a group with an unsolvable word problem. Consider a homology sphere constructed by the method of the previous section with fundamental group the universal central extension of A_w, the group whose triviality would witness the triviality of w in A. Let us call this manifold $\Sigma(w)$. Note that we have arranged that $\Sigma(w)$ is the sphere iff w is the trivial word. If w is not trivial, then $\Sigma(w)$ is not even simply connected.

Now let M be an arbitrary manifold. Consider $M\#\Sigma(w)$. If $\Sigma(w)$ is the sphere, this manifold coincides with M (up to diffeomorphism). However, if we are in the w nontrivial case, then $M\#\Sigma(w)$ has a different fundamental group from M, because of the following group theoretic lemma:

Lemma *If G and K are finitely generated groups such that G is isomorphic to $G * K$, then K is trivial.*

This follows from Grushko's theorem, which asserts that any surjection from a free group to a free product can be decomposed as the composite of an automorphism of the free group and a free product of surjections. (Note, by the way, that this argument also shows that no group is algorithmically recognizable; this is not a Markov property!)

Remark. Here is an alternative to any theory of universal central extensions, which gains us explicitness. If π is a perfect group with

$$\pi = \langle g_1, \ldots, g_k \mid r_1, \ldots, r_m \rangle,$$

then we can form the group

$$E = \langle g_1, \ldots, g_k, t_1, \ldots, t_m \mid g_1 = w_1, \ldots,$$
$$g_k = w_k, t_1 = r_1, \ldots, t_m = r_m, [g_a, t_b] = e \rangle$$

where the w's are explicit products of commutators of the g's. (They should be relations that actually hold in π.) Note that E is a central extension of π, so E is nontrivial if π is. If π is trivial, then the t's are all central, but since they equal the r's they normally generate the group, so the group is commutative. One quickly sees then that the g's generate, but since they are commutators, the group is trivial.

Now, using the formula $H_2(G) = [F : R]/[R : R]$ (Hopf) in terms of an isomorphism $G = F/R$, where F is free, one sees that by definition $H_2(E) = 0$.

Using the explicit rewriting of elements of $[F : R]$ in terms of $[R : R]$, one can give a quite direct constructive proof of the Hopf exact sequence and explicitly produce the necessary 2-sphere. The proof of this is not much harder than an explicit proof of the Hurewicz isomorphism theorem in dimension two; to replace a 2-cycle by a sphere, one just surgers (nullhomotopic) curves on the cycle to get the desired bound. (See figure 9 for the simply connected case.)

The upshot of these explicit formulas is that one can actually produce relatively "small" or "simple" "witness homology spheres" from the data of G and w. For instance, we could get an upper bound on the number of simplices we would need in $\Sigma(w)$ just from knowledge of the length and complexity of the defining relations of G and the word w. These considerations will gain relevance in the later parts of the book.

For future reference, we will now prove a number of other similar theorems.

Theorem *There is no algorithm to decide whether or not an embedding of S^n in S^{n+2} is knotted if $n > 2$.*

Theorem *There is no algorithm to tell whether or not an embedding of $S^1 \times S^n$ in S^{n+2} is isotopic to the standard embedding if $n > 2$.*

Remark. An extension of this will be given in appendix 2 to section 2.4.

Recall that for the sphere there is in fact a unique embedding, up to reparametrization, so the algorithm is rather dull! For arbitrary simply connected hypersurfaces, there is an algorithm, despite the fact that the set of isotopy classes of embeddings can well be infinite (even up to reparametrization).

Proofs. We will prove the theorem for knots; the result about $S^1 \times S^n$ follows, because there is a canonical $S^1 \times S^n$ near any knot, and the knots are isotopic iff the hypersurfaces are.

Our knots are essentially of the following sort: consider $S^1 \times \Sigma(w)$. We shall find a curve γ in $S^1 \times \Sigma(w)$ which, when we surger, will give us S^{n+2}.

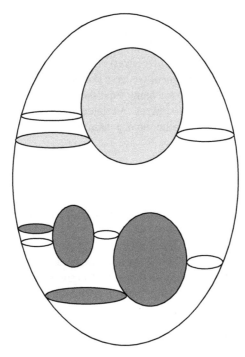

Figure 9. The Hurewiz theorem replaces a surface (e.g., 2-cycle) with a 2-sphere in a
simply connected space, essentially by filling in the 1-cycles of the surface
by disks. A similar construction can be done whenever one has an *explicit*
trivialization of the second homology of the fundamental group.

Recall that in the surgery process the circle is replaced by an S^{n+2}; this is the
knot; it will bound a punctured $\Sigma(w)$ in S^{n+2}, so if w is trivial, it will be the
unknot. On the other hand, the fundamental group of the complement of this
knot is $\mathbb{Z} \times \pi_1(\Sigma(w))$, so if w is not trivial, the knot is nontrivial.

Our curve γ is designed simply to represent the element $(1, w)$ in $\mathbb{Z} \times
\pi_1(\Sigma(w))$. One readily checks that the result of the suggested surgery is a
homology sphere, whose fundamental group is $\mathbb{Z} \times \pi_1(\Sigma(w))/\langle(1, w)\rangle$, which
is exactly $\pi_1(\Sigma(w))/\langle[w, \pi_1(\Sigma(w))]\rangle$. So one has to check that one can construct the witness so that the witness group is made trivial not only by killing
w but even by making it central. This is not hard to arrange. (See the notes.)

Remark.   A great deal of what will be done throughout the remainder of this
book is producing witnesses with given strong propertics. (We call these dichotomy theorems.) Geometric witnesses are another term for what we have
been calling doppelgangers.

Remark.   The status of the results of this section is unsettled in dimension four. One does know Markov's theorem, which asserts that there is no algorithm that can determine whether 4-manifolds are homotopy equivalent, or homeomorphic, or diffeomorphic. But it is still conceivable that there is an algorithm to recognize the 4-sphere. The argument we have sketched is irremediable, because, as we mentioned before, there are many superperfect groups that are not fundamental groups of homology 4-spheres. A related unsolved problem is "Is there an algorithm to determine whether an n-generator, n-relation group is trivial?"

2.3 NABUTOVSKY'S THESIS

We shall now give our first application of the main theme of this book and prove a theorem of Nabutovsky's thesis.

Recall that we saw that, as a consequence of Smale's theorem, there is a unique embedding of S^n in S^{n+1} up to reparametrization and isotopy. Having thus learnt that all these embeddings can be "straightened," our interest now turns to the question of how complicated a fairly simple embedding must get on its way to straightening out.

Definition.   A hypersurface Σ has *crumbledness k* if the largest size embedded ribbon around it has radius $1/k$ (i.e., the map $s + t\nu: \Sigma \times [-r, r] \to S^{n+1}$, where ν is the unit normal to Σ at s is a diffeomorphism for r smaller than $1/k$).

Theorem   *There is no computable function f such that any hypersphere in S^n, $n > 5$, with crumbledness less than k can always be isotoped back to the equatorial embedding through embeddings with crumbledness less than $f(k)$.*

Later, we will give more precise statements than this one; for instance, the rate of necessary growth of crumbledness grows like the busy beaver function, and thus exceeds any computable function on all sufficiently large inputs. (The same reasoning applies to arbitrary hypersurfaces.)

The idea of the proof goes like this. First, one proves a "compactness theorem" for the space of embeddings of all manifolds with crumbledness less than k. (This has to be a fairly effective proof.)

Now suppose that one knew that, for all k, for hyperspheres one could always isotope a given crumbledeness k embedding to the round sphere. We can use the results of section 2.1 to produce a sequence of homology spheres in the sphere, whose crumbledness can be measured. Now, check if they can be isotoped to the round sphere by paths through embedding space with crumbledness less than $f(k)$; this is a computable task by the computability of f and the compactness theorem. If the homology sphere were the sphere, we would succeed (by assumption), and, of course, if it weren't, we couldn't. Since no algorithm exists, Nabutovsky deduces the theorem.

Note that the argument also shows that, for an infinite sequence of integers k, the set crumbledness $^{-1}[0, k]$ is disconnected, and contains components where no point has crumbledness less than $k - 1$. One could expect that by combining compactness with these components, one would get an infinite number of local minima of the crumbledness functional. Nabutovsky calls these "self-clenching thick hyperspheres" and shows that an infinite number of these exist, by exactly those steps.

Theorem *There are infinitely many local minima of crumbledness: $Emb(S^n \subset S^{n+1}) \to \mathbb{R}$; these can be taken to have smoothness of class $C^{1,1}$.*

Again, we note that, since $Emb(S^n \subset S^{n+1})$ is connected, one feels entitled on general grounds to at most one local minimum. It is the encoding of an unsolvable problem into this functional that guarantees many more.

We shall give more details of these kinds of arguments in chapters 3 and 4. Essentially one shows how to approximate hypersurfaces by zeros of polynomials with bounds on degree and coefficients. Then one is in the setting of real algebraic geometry and one can make good use of the Tarski-Seidenberg theorem and related triangulation theorems in constructing algorithms.

Also remark that, if one were interested in nonsimply connected hypersurfaces, one need not use zeros of polynomials, but could instead use polynomial functions more directly, and the analogous compactness results are much simpler and probably could be left as an exercise. Using this method one could study the distortion of embeddings $Dist: Emb(\Sigma \subset S^{n+1}) \to \mathbb{R}$, defined by

$$Dist(f) = \sup(d(f(x), f(y))/d(x, y), d(x, y)/d(f(x), f(y)) + ||f||_{C^2},$$

where x, y lie in the hypersurface (and d denotes distance in Σ or S^{n+1}, depending on the context); note that this function does depend critically on the parametrization. (We have added the C^2-norm of f to the most straightforward definition of distortion to guarantee that only smooth embeddings have finite distortion—but I believe that the same results are true without this term, because of an approximation theorem for topological embeddings due to Ancel Cannon.) For this functional the above theorem is false for simply connected hypersurfaces, but using the results of appendix 2 to following section 2.4, one obtains the analogous theorem to Nabutovsky's for this functional for nonsimply connected hypersurfaces.

2.4 THE CLASSIFICATION OF HOMOLOGY SPHERES

For deeper applications it is not enough to have doppelgangers that are nontrivial only because they are nonsimply connected. For instance, to show that there is no algorithm to decide whether a manifold has a metric of positive scalar curvature (which arises in one approach to the problem of the construction of Einstein metrics) we cannot use the homology spheres constructed so far.

Indeed, the construction we gave actually produces homology spheres that always have positive scalar curvature.

In this section I will explain some very nice work of Hausmann and Vogel on the classification of homology spheres. In appendix 1 we will give a more detailed exposition of this work, and in appendix 2 we will apply this to the classification of hypersurfaces in the sphere, and to give a version of Nabutovsky's theorem from the previous section where one need not mod out by the diffeomorphism group.

First, I should emphasize that there is no known theory of classification of homology spheres up to homeomorphism or diffeomorphism, with just a given fundamental group. It is somewhat akin to trying to classify knots up to isotopy, which has been done only in "a stable range"; perhaps there is a "stable range" theory of homology spheres which would be useful to us. But what we will discuss here is much cruder.

Definition. $\theta^n(\pi)$ is the group whose generators are oriented n-dimensional homology spheres Σ, equipped with a map $\pi_1(\Sigma) \to \pi$. Addition is defined by the connected sum, and we will view as trivial any homology sphere which bounds an acyclic manifold, such that the map of the fundamental group to π extends. (This group is different in the smooth and PL categories, because of exotic differential structures on the sphere.)

The following result of Hausmann computes $\theta_n(\pi)$ for $n > 4$.

Theorem *For $n > 4$, and π a finitely presented superperfect group, we have $\theta^n(\pi) = \pi_n(B\pi^+) \oplus \theta^n$, where θ^n denotes $\theta^n(e)$ (which vanishes in PL and Top, and is the group of smooth structures on the n-sphere for the smooth category), and $+$ denotes the Quillen $+$ construction (reviewed below). All elements are represented by homology spheres with fundamental group exactly π.*

Now for the $+$ construction.

Proposition/Definition

(Quillen.) For any X and any perfect normal subgroup P of $\pi_1(X)$, there is a unique space X^+ equipped with a map $X \to X^+$ such that

1. $\pi_1(X^+) = \pi_1(X)/P$; and
2. $H_k(X) \to H_k(X^+)$ is an isomorphism for all (local) coefficient systems defined over X^+.

It is natural. (If one has a map $X \to Y$ and corresponding subgroups of the fundamental group, then there is a map $X^+ \to Y^+$.) The uniqueness follows from obstruction theory. (Two candidates for X^+ would have natural maps to each other, which one then checks have to be homotopy equivalences.)

The construction of X^+ goes like this. Let $p_1, p_2, \ldots$ be generators for P. Attach 2-cells to X to kill P. Unfortunately, we have created some two-dimensional homology by this process (indeed, one generator for each of the p's). One can check, and in our case this is obvious because we are killing the whole fundamental group, that these new generators are in the image of the Hurewicz map from π_2; one attaches a 3-cell along each of these generators, and one readily sees that (1) and (2) hold. The naturality is clear from the description; it also follows from obstruction theory and the homological characterization.

I will describe the ingredients and sketch the proof of this theorem in appendix 1; I believe that it and its extensions are likely to be useful in other contexts.

The map $\theta^n(\pi) \to \pi_n(B\pi^+)$ is defined at the level of representative elements by the induced map on + constructions $\Sigma \to B\pi$. (Of course, Σ^+ is homotopy equivalent to the sphere, since the degree-one map $\Sigma \to S^n$ satisfies conditions 1 and 2 of the definition.)

Corollary 1 *For π (super)perfect, an element of $H_n(B\pi)$ can be represented by a cycle which is a homology sphere iff its image in (the isomorphic group) $H_n(B\pi^+)$ is in the image of the Hurewicz homomorphism $\pi_n(B\pi^+) \to H_n(B\pi^+)$.*

This follows directly from the classification theorem.

Remark. Since $B\pi^+$ is simply connected, in practice, the main obstacle to computing the Hurewicz homomorphism is actually finding out its homology. For instance, one cannot tell whether or not a group is superperfect (just consider $A *_{\langle g \rangle} A$, where A is an acyclic universal group; this has $H_2 = 0$ iff g is trivial; otherwise it has $H_2 = \mathbb{Z}_n$, where $\langle g \rangle$ is a cyclic group of order n.)

As a simple corollary, we have

Corollary 2 *There are homology spheres with no metric of positive scalar curvature. In fact, there is no algorithm to decide whether or not a homology sphere has a metric of positive scalar curvature.*

Proof. The first statement follows from the corollary and known results about the Novikov conjecture that will be reviewed in appendix 3 of this section. These results imply that for many torsion-free groups π a homology sphere Σ with fundamental group π has a metric of positive scalar curvature iff the associated smooth homotopy sphere does, and $[\Sigma]$ in $H_n(B\pi)$ is trivial.

To prove the algorithmic unsolvability of the search for positive scalar curvature metrics one can proceed in two different ways. First, one can do a witness construction to produce a homology sphere that is either the sphere or represents a nontrivial element in $H_n(B\pi)$ if π is nontrivial. A second way would be to produce a group for which $\pi_n(B\pi^+) \to H_n(B\pi^+)$ is onto (even an isomorphism, if one wishes) so that the whole homology is represented by homology spheres, but whose homology is a c.e. noncomputable abelian group.

Appendix 1: Surgery, Homology Surgery, and All That

This appendix will give a rapid summary of some of the formal apparatus used for the classification of manifolds, knots, and homology spheres in high dimensions. The first critical theorem in this development is the s-cobordism theorem (which generalizes the theorem of Smale that opened section 2.1).

Definition. An h-cobordism on M^n, a compact manifold, is a compact W^{n+1} with two boundary components $\partial W = M \cup M'$, and both M and M' include into W as deformation retracts.

Theorem *(s-cobordism theorem) The set of h-cobordisms on M^n with $n > 4$ is in a 1-1 correspondence with an abelian group $Wh(\pi_1(M))$.*

This theorem is true in all of the usual categories of geometric topology: Diff, PL, and Top. There are also noncompact and equivariant versions, and many other generalizations. It is called the s-cobordism theorem, because it asserts that if a certain invariant (called the torsion) of the inclusion map $M \to W$ vanishes (is a simple homotopy equivalence) then $W = M \times [0, 1]$.

The group $Wh(\pi)$ is purely algebraically defined. It is $\lim H_1(GL_n(\mathbb{Z}\pi))/ \pm\pi$, where $\mathbb{Z}\pi$ is the integral group ring, GL_n denotes the general linear group of invertible matrices (GL_n is included in GL_{n+1} by taking the direct sum of an automorphism with the identity), and $\pm\pi$ consists of the obvious invertible 1×1 matrices.

Together with a little algebra, the theorem can be used to construct some very interesting manifolds, but its impact is often the reverse. It gives a method for showing that homotopy equivalent manifolds are isomorphic. Find an h-cobordism between them, and see that its torsion is trivial.

Remark. In all the known calculations for π torsion-free, the group $Wh(\pi)$ vanishes. We shall discuss Whitehead theory more in chapter 4.

Surgery theory provides a technique for implementing this program. We will describe this only in the situation of compact closed manifolds. (There are versions for manifolds with a boundary, noncompact manifolds, homology manifolds, stratified spaces, group actions, Poincaré spaces, etc.) The idea is that homotopy theory provides one with good tools to tell if two manifolds are cobordant, so one wants to have a procedure for turning a cobordism into an s-cobordism.

Actually, the way it works is like this. One starts off with a Poincaré complex X, that is, a space that satisfies a strong form of Poincaré duality and a degree-one map $f: N \to X$ that satisfies some additonal condition on tangent bundles.

The goal is to find a cobordism (by the technique of "surgery") of f to a homotopy equivalence. The obstruction to success lies in an "L-group."

Theorem   *The following sequence is exact for M an n-manifold:*
$$\cdots \to [\Sigma M : F/Cat] \to L_{n+1}(\pi_1(M)) \to S(M)$$
$$\to [M : F/Cat] \to L_n(\pi_1(M)).$$
Here $S(M) = \{(M', f) \mid f : M' \to M \text{ is a homotopy equivalence }\}/ h\text{-cobordism}.$

F/Cat is a classifying space, whose homotopy type has been much studied; it is completely understood in the PL and Top settings, and is (always) rationally the product of Eilenberg-MacLane spaces. Also $[M : F/Cat]$ is rationally $\oplus H_{4i}(M)$, which measures the difference in the Pontrjagin classes of M and M'. The L-groups are Grothendieck groups of quadratic forms or their automorphisms. They are 4-periodic, and like Wh, purely algebraically defined. The groups $[M : F/Cat]$ measure these cobordism groups.

Notice that $S(M)$ is only a set with distinguished element (id: $M \to M$). The "map" $L_{n+1}(\pi_1(M)) \to S(M)$ is actually a group action of $L_{n+1}(\pi_1(M))$ on the set S. The "kernel" is the isotropy of the distinguished element. The meaning of the sequence is that one tries first to see if two "structures" on M (or a Poincaré duality space X) are cobordant. That is what the map to $[M : F/Cat]$ measures. Not every cobordism class contains such a representative; that is where L_n enters.

Moreover, when two "structures" are cobordant, we can try to cobord the cobordism to a homotopy equivalence (to $M \times [0, 1]$). If we succeed, they are h-cobordant. If we fail, there is still one more thing to try: maybe a different cobordism will have vanishing surgery obstruction. That is the meaning of the map $[\Sigma M : F/Cat] \to L_{n+1}(\pi_1(M))$.

Remark.   In fact, from $L_{n+1}(\pi_1(M))$ to the left, the surgery sequence becomes a sequence of abelian groups and homomorphisms. Remarkably, in PL and Top, the set $S(M)$ can be given the structure of an abelian group, and with the right H-space structure on F/Cat, the whole sequence becomes a sequence of groups and homomorphisms. (In the smooth category, it turns out that no such H-space structure exists.)

It is in the topological category that the surgery exact sequence takes its most elegant form.[46] In particular $[M : F/Cat]$ Poincaré-dualizes to a homology theory, denoted by $H_n(M; \mathbb{L})$, and the sequence becomes a covariantly functorial 4-periodic exact sequence abelian group:
$$\cdots \to H_n(M; \mathbb{L}) \to L_{n+1}(\pi_1(M)) \to S(M) \to H_n(M; \mathbb{L}) \to L_n(\pi_1(M)).$$
Naturality implies that $H_n(M; \mathbb{L}) \to L_n(\pi_1(M))$ actually factors through a map $H_n(B\pi_1(M); \mathbb{L}) \to L_n(\pi_1(M))$, called the *assembly map*. We will discuss it more in appendix 3 on the Novikov conjecture.

---

[46]There is a slight lie here that can be corrected with homology manifolds. See the references.

Remark. The analogue of the conjectural vanishing of Whitehead groups for torsion-free groups is that the assembly map should be an isomorphism for torsion-free groups. Again, there have been many calculations that bear this out.

Surgery theoretic ideas can not only be used to classify manifolds up to h-cobordism or isomorphism, they can also be modified to give results about concordance classifications in a homological sense. Examples include the results about homology spheres we discussed and the classification of knots up to cobordism. (A knot of S^n in S^{n+2} is cobordant to another one if together they bound an $S^n \times [0, 1]$ in $S^{n+2} \times [0, 1]$.)

Recall that the L-groups measured the obstruction to taking a cobordism with a map to M, and making this map into a homotopy equivalence. Γ-groups are defined to measure similar obstructions to making a map from a space with fundamental group π into an R-homology equivalence, where R is some ring. (There are technical conditions on the map $\mathbb{Z}\pi \to R$, but those are always satisfied for surjections. If R has an augmentation to $\mathbb{Z}$, then the "homological" part, i.e., the normal cobordism set part, of the classification is as in the usual surgery theory.) The notation for these groups is $\Gamma_n(\mathbb{Z}\pi \to R)$. Note that $\Gamma_n(\mathbb{Z}\pi \to \mathbb{Z}\pi) = L_n(\mathbb{Z}\pi) = L_n(\pi_1)$, because a $\mathbb{Z}\pi$ homology equivalence between spaces with fundamental group π is a homotopy equivalence. Knot theory is studied by an analysis of a surgery exact sequence with a good deal of pressure put on the determination of the group $\Gamma_n(\mathbb{Z}[\mathbb{Z}] \to \mathbb{Z})$.

The results about homology spheres discussed in this section follow from a homotopy theoretic analysis (see below) combined with the following calculation:

Theorem $\Gamma_n(\mathbb{Z}\pi \to \mathbb{Z}\pi') = L_n(\pi')$ *if $\pi \to \pi'$ is a surjection with perfect kernel.*

The preceding theorem and the following analysis can be used to analyze the aggregate of manifolds homology equivalent to a given one, with larger fundamental group.

Now, let us analyze $\theta^n(\pi)$. We already discussed the map $\theta^n(\pi) \to \pi_n$ $(B\pi^+)$. Consider the homotopy pullback

where the vertical arrow is determined by an element of $\pi_n(B\pi^+)$. This will give us a space X with fundamental group π, and with the integral homology of the sphere. These X's are suitable for homology surgery, just as genuine Poincaré spaces were used in ordinary surgery.

If one takes a degree-one map $M \to X$, then one can try to turn it into a homology equivalence, with an obstruction in $\Gamma_n(\mathbb{Z}\pi \to \mathbb{Z}) = L_n(e)$. However, using a simple naturality comparing X to S^n one can arrange $M \to S^n$ and therefore $M \to X$ to have vanishing simply connected surgery obstruction, and one gets a manifold integrally homology equivalent to X with fundamental group mapping to π.

In other words, the two parts of the classification theorem are somewhat different: the $\pi_n(B\pi^+)$ parametrizes a class of homotopy types, and the θ^n is what classical surgery theory asserts about the classification within a homotopy type for X^+ (for any of these X's).

Remark. (For simplicity) if M is simply connected, then the homology h-cobordism classes of manifolds homology equivalent to M and the superperfect fundamental group π are in a 1-1 correspondence with $S(M) \times [M : B\pi^+]$.

Appendix 2: Isotopy of Hypersurfaces

Our goal is the following:

Theorem *Let M^n be a nonsimply connected compact manifold whose first homology is trivial and which embeds in S^{n+1}. Then there is no algorithm to determine whether any other embedding is isotopic to the given one. (In particular, such an M has infinitely many smooth embeddings.)*

As we remarked in section 2.2, there is an algorithm for the simply connected case. The infinitude of embeddings can be proven quite easily by classical methods related to the Alexander polynomial if H_1 is nontrivial. The techniques of this appendix can be extended to the general nonsimply connected case, but the details are somewhat more cumbersome, so I shall avoid them.

Let M be a hypersurface. We will say that M is *simple* if the fundamental group of each of the two components is trivial.[47]

Proposition *If $n > 3$, a hypersurface in S^{n+1} has a simple embedding iff $H_1(M) = 0$.*

The necessity follows from Mayer-Vietoris. The sufficiency is an application of Alexander duality and a geometric plus construction. (Note that, if one is given a hypersurface, it need not be simply embedded; the proposition merely asserts that there is an associated simple re-embedding of the hypersurface.)

It turns out that the simple embeddings can be classified algorithmically and can be unique. (The simple embeddings of M are in a 1-1 correspondence

---

[47] One could redo all of the succeeding arguments using hypersurfaces whose complements have a component with perfect fundamental group and focusing attention on that component. However, with only this condition, one cannot give a clean classification as far as I know.

with the embeddings of M^+.) Our goal is to show that there are always many nonsimple embeddings.

Proposition 1 *For any group π and any nontrivial w in π, there is a group G_w, so that π embeds in G_w, the induced map of the embedding $\pi \rightarrow G_w$ in homology is trivial, G_w is superperfect, and w normally generates G_w. Moreover, this is true in the dichotomy sense: if w is a word, then G_w can be algorithmically constructed to have these properties, with the exception that if w is trivial, so is G_w.*

This can be achieved as an amalgam of several constructions that we have already encountered. We first embed π into a universal acyclic group A (see [BDM]), and then perform the Adian-Rabin construction (see chapter 1, section 1.2) on A with w, and take its universal central extension (as in the proof of Novikov's theorem). This group has all of the desired properties.

Proposition 2 *For any nontrivial group π, there is a c.e. sequence of superperfect groups G_j, $j = 1, 2, 3, \ldots$, such that, for each i, there is a map $\pi \rightarrow G_j$ which is either an injection or trivial, but the image always normally generates and is trivial on group homology. Moreover, the set of j for which G_j is trivial is a c.e. noncomputable set.*

This can be done in two steps. One uses the previous proposition applied to $\pi * B$ for any group B with unsolvable word problem, first with respect to a nontrivial element of π and then to the sequence of words of B.

Remark. There is a stronger condition on a homomorphism between groups than triviality on homology: it is *stable triviality*, that is, that the map between classifying spaces is nullhomotopic after suspension.

Addendum

In the previous propositions, one can assume stable triviality of the inclusion.

The constructions do not have to be changed!

Now the proof of the theorem follows. We use homology surgery to construct embeddings of M where one complementary region is simply connected, and the other has fundamental group G_w, and whose + construction was some originally given embedding. This embedding is isotopic to the original embedding iff w is trivial.

In more detail (for those with some experience with surgery theory), since M is trivial in framed cobordism of the trivial group (it bounds a region in the sphere!), it bounds in $\Omega^{fr}(G_w)$. In fact, it is trivial in $\Omega^{fr}(X \times G_w)$, where X is one of the complementary regions. This gives a manifold with fundamental group $\Omega^{fr}(\pi_w)$ with a map to X, that one can hope to homology surger. The theorem from appendix 1 implies that the surgery obstruction vanishes in odd dimensions. In even dimensions there is a final simply connected surgery obstruction that one has to worry about, but this can be avoided in the PL category by taking the connected sum with a Kervaire or Milnor manifold, and finally applying smoothing theory to return to the smooth category.

Appendix 3: The Novikov Conjecture

It is possible to write endlessly about the Novikov conjecture. However we will try to be as brief as possible and refer the reader to the references. The original Novikov conjecture is the statement that a certain combination of characteristic classes is a homotopy invariant for nonsimply connected manifolds.

Conjecture *If $f : M \to B\pi$ is a map, then $f_*(L(M) \cap [M]) \in H_*(B\pi; \mathbb{Q})$ is an oriented homotopy invariant.*

Here $L(M)$ is the Hirzebruch polynomial in the Pontrjagin classes. It is characterized by $L(M \times N) = L(M) \times L(N)$ and the condition that $\langle L(\mathbb{CP}^{2k}), [\mathbb{CP}^{2k}] \rangle = 1$. Hirzebruch's signature formula is the assertion that for π the trivial group this is correct, and the invariant is, in fact, the signature of the cup product pairing on $H^{2k}(M; \mathbb{Q})$.

The Novikov conjecture then asserts that in the nonsimply connected case there are additional combinations of Pontrjagin classes that are homotopy invariant; since these are bordism invariant, surgery theory implies that such a statement must be reflected in the L-groups. This leads to the following (equivalent) formulation.

Conjecture *The assembly map (see appendix 1) $A: H_n(B\pi_1(M); \mathbb{L}) \to L_n(\pi_1(M))$ is rationally injective.*

Indeed, for torsion-free groups one has the following.

Borel Conjecture

The assembly map for π torsion-free, given by $A: H_n(B\pi_1(M); \mathbb{L}) \to L_n(\pi_1(M))$, is an isomorphism.

Note

There is even a version for groups with torsion that explicitly takes the finite subgroups into account, but it is a bit too complicated to discuss here. We will return to it in the final chapter.

It is now known that the Borel conjecture is correct for torsion-free discrete subgroups of linear real Lie groups and the Novikov conjecture is true for all linear groups.

Another important inductive method is based on the following Mayer-Vietoris sequence for L-groups (ignoring some difficulties at the prime 2):

$$\cdots \to L_k(B) \to L_k(A) \times L_k(C) \to L_k(A *_B C) \to L_{k-1}(B) \to \cdots$$

and similarly for HNN extensions. Since group homology (even with coefficients in a spectrum) satisfies Mayer-Vietoris, one can inductively use these sequences and the 5-lemma to get new cases of the Borel conjecture from old. It takes some more care to combine cases of the Novikov conjecture, but this is sometimes (worthwhile and) doable.

These remarks imply that we can use assume that we know that the Borel conjecture holds for many of the groups arising in chapter 1. The major exceptions are results that depend on the use of a universal group. (Universal groups are never torsion-free; it is never the case that the usual assembly map is rationally an isomorphism.) However, the careful reader can check that, having delayed use of the universal group till the very end, all of the strange groups that have arisen so far actually do satisfy the Novikov conjecture.

Remark. There are analogues of these conjectures (and known cases) in algebraic K-theory, and also in the K-theory of operator algebras. Special cases of both of these have geometric applications; the operator theoretic version, applied to the "signature operator" implies the usual Novikov conjecture. It has the advantage of applying to other operators as well. Another consequence of this conjecture is that if a spin manifold defines a nontrivial rational class in $H_n(B\pi; \mathbb{Q})$ then it cannot admit a metric of positive scalar curvature.[48]

2.5 SIMPLICIAL NORM

As part of our program of obtaining results about variational problems, we will need results that prevent minimizing sequences for our functionals from "falling off the space." Analytically, this is often accomplished by a priori inequalities. The theory of the simplicial norm is one method for getting a priori bounds on the volume of a manifold.

Definition. Let M be a topological space. Give $C_*(M)$, the real singular chain complex of M, the L^1-norm. This induces a pseudonorm $|\cdot|$ on $H_*(M; \mathbb{R})$; a homology class is assigned the infimum of the norms of all singular chains representing that class.

Note that $[S^n] = 0$ for all $n > 0$. We can represent k times the fundamental class using just one n-simplex that goes around the sphere k times. Letting k go to infinity gives us the desired vanishing. In fact, this norm is trivial on all simply connected spaces. For general X, the map $H_*(X) \to H_*(B\pi_1(X))$ is

---

[48]This is a special case of a more general statement asserting that the "higher A-genus" obstructs positive scalar curvature.

a pseudo-isometric embedding; the only elements that die have norm 0. It is quite nontrivial that this norm is nontrivial:

Theorem 1   *If M is a negatively curved Riemannian manifold, and x is a nontrivial element of $H_k(M)$ with $k > 1$, then $|x| > 0$.*

The proofs of this and many other aspects of the theory are facilitated by introducing the dual theory *bounded cohomology*. In particular, $H_b^*(X)$ is the cohomology of the cochain complex of bounded cochains on X. A bounded k-cochain is one that is uniformly bounded on all k-simplices. Nonexamples are provided by integration of forms. The integral of the volume form of the singular simplex that goes around the sphere k times is k times the volume of the sphere. The proof of the theorem comes about by finding a bounded cohomology class that pairs nontrivially with the class x.

Hahn-Banach remark

We have $|x| > 0$ iff there is a bounded cocycle c such that $c(x)$ is nonzero.

Now, here is the relevant bounded cocycle. Given a simplex, lift it to the universal cover of M and ignore everything about this lift except where its vertices go! Now draw the geodesic from the first vertex to the second, all the lines from this arc to the third, and so on, repeatedly coning. This is called the *straightened simplex*. Let w be a closed k-form which detects x. Our cochain now assigns to a simplex the integral of w over the straightened simplex. It is a fact about negatively curved manifolds that, for $k > 1$, there is a universal bound on the volume of simplices.[49] Thus this cochain is bounded and it detects x, as advertised, completing the proof.

Amenable groups play a special role in the theory of the simplicial norm. A group G is amenable if it possesses a G-invariant mean on $\ell^\infty(G)$. An example would be a weak-star limit of $1/(2n + 1)(a_{-n} + a_{-n+1} + \cdots + a_{n-1} + a_n)$ for $G = \mathbb{Z}$. All abelian groups (indeed all solvable groups) are amenable by a similar obvious (or not so obvious) mean. The simplest nonamenable group is the free group.

Proposition   *If G is amenable, then $| \cdot |$ is trivial on $H_*(BG)$.*

Proof. It is a little easier to use cohomology, that is, to show the vanishing of the bounded cohomology $H_b(BG)$. The averaging process on G directly gives rise to an "averaging" chain map[50] of $C_b^*(EG)$ into $C_b^*(BG) = C_b^*(EG)^G$ which is norm nonincreasing. Moreover, the obvious map $C_b^*(BG) \to C_b^*(EG) \to C_b^*(EG)^G$ is the identity. So one obtains that the map $H_b^*(BG) \to H_b^*(EG)$ is an isometric injection. Since the latter group vanishes, the result follows.

---

[49]This follows from comparison theorems and a bit of hyperbolic geometry.
[50]Recall that EA is the universal cover of BA.

More generally one has the following, and these require more work:

Theorem 2   *If $G = K *_A L$ where A is amenable, then the map $H_a(K) \times H_a(L) \to H_a(K *_A L)$ is a pseudoisometric embedding (again, norm-zero elements may go to zero).*

Theorem 3   *If $G \to K$ is a surjection with amenable kernel, then $H_k(G) \to H_k(K)$ is a pseudoisometric embedding (and, as before, norm-zero elements may go to zero).*

The second theorem can help us when we consider universal central extensions; the first will require us to modify our Adian-Rabin constructions to go away from free groups and toward amenable groups.

2.6 HOMOLOGY SPHERES WITH NONZERO SIMPLICAL NORM

Theorem 1   *In all dimensions $n > 4$, there are homology spheres Σ with nonzero simplicial norm.*

Remark.   As a consequence, one can obtain homology spheres with arbitrarily large simplicial norms, merely by taking many connected sums of Σ with itself. (Hint: use theorem 2 of section 2.5).

The proof of this theorem starts with Clozel's theorem about the cohomology of appropriate lattices in $U(k, 1)$ explained in chapter 1, section I.8. Noting that these arithmetic varieties (which are the $B\Gamma$'s) are complex hyperbolic manifolds, so that theorem 1 above applies, we would like to get homology spheres that represent nontrivial homology classes in this manifold. (We are not asserting that there are homology spheres among these manifolds: there surely are not! And, besides, they are all even dimensional.)

There are a few fairly minor difficulties. First, one knows only the betti numbers of these varieties, not their homology. Second, they are not superperfect: the Kahler (= hyperplane) class is nontrivial, so it will be important to take universal central extensions to get rid of this. And, finally, we will have to check that the criterion for representing homology classes by homology spheres actually holds. As usual, we will also find the following sort of result useful:

Theorem 2   *For each $n > 4$, there is a Turing machine which outputs homology spheres Σ_k for $k = 1, 2, 3, 4, \ldots$ such that, for each k, either $\Sigma_k = S^n$ or $|[\Sigma_k]| > 1000$. Moreover, the set $\{k \mid \Sigma_k = S^n\}$ is not computable. More informally, there is no method to distinguish the sphere from manifolds that otherwise have a large simplicial norm.*

Both theorems are proven by essentially the same method. Let us begin with theorem 1. Let Γ be one of the groups that Clozel studied for $U(2n + 1, 1)$ (see chapter 1, section I.8); without loss of generality Γ can be assumed torsion-free. Nothing forces $H_1(\Gamma)$ to vanish. (Indeed, there are lattices satisfying his conditions for which this fails, and I do not know whether there are any for which it is true.) Let $g_1, g_2, \ldots, g_k$ be generators, and consider $\Delta = \Gamma *_{\mathbb{Z}} A *_{\mathbb{Z}} \cdots *_{\mathbb{Z}} A$, where A is torsion-free acyclic and each successive amalgamation is along a different generator of Γ.

Now Δ is perfect, but not superperfect. So now we take the $\mathbb{Z}^{k+1}$ central extension Ω (= universal central extension) of Δ and it is now superperfect. We claim that there is an element in $H_n(BG\Gamma^+)$ that has a common image in $H_n(B\Delta^+)$ as an element of $H_n(B\Omega^+)$, which lies in the image of the Hurewicz homomorphism for all of these spaces. Theorems 1, 2, and 3 and corollary 2.2 now combine to give us our desired homology spheres.

In order to do this calculation, we shall use a slight modification of the + construction, which applies whenever H_1 is a torsion group. One follows exactly the process used in the + construction, using generators for π_1 and then killing the excess H_2 to the best extent one can, and finally rationalizing.[51] It has a similar homological characterization to the + construction.

Now, we know the homology groups of $BG^+_{\mathbb{Q}}$. They are given by Clozel's theorem. We also know what is the homology of the circle bundle Ψ over it, associated with the Kähler class: By the Gysin sequence (see chapter 1, section I.7),[52] its first homology group arises in dimension n; by the Hurewicz theorem, this lies in the image of the Hurewicz homomorphism. By naturality, these elements can be pushed forward to $\pi_n(B\Delta^+_{\mathbb{Q}})$ and $H_n(B\Delta^+_{\mathbb{Q}})$, but since $B\Delta^+$ is simply connected, we can pull back (a finite multiple of) these classes to $B\Delta^+$.

Note that $B\Delta^+_{\mathbb{Q}}$ resembles $B\Gamma^+_{\mathbb{Q}}$ wedged with k rationalized copies of S^2. Thus one can easily compare the Mayer-Vietoris sequence for computing the nth homology of $B\Omega^+$ as being that of $T^k \times \Psi$ together with $S^1 \times$ (a T^k-bundle over a wedge of 2-spheres) along a copy of T^{k+1}. The key point is that, since the gluing is over T^{k+1}, there is no room to kill the element coming from $1 \times \Psi$. Thus, our element survives; it is in the image of the Hurewicz, and has nonzero norm.

Serre's mod C Hurewicz theorem asserts that these rational classes pull back (perhaps after multiplying them by large integers) to classes in $\pi_n(B\Omega^+)$ and $H_n(B\Omega^+)$. This completes the proof of theorem 1.

To prove theorem 2, we just have to take the homology sphere from theorem 1 and marry it to an appropriate witness construction. Let G_w be a witness group to the triviality of a word w in a torsion-free group with unsolvable problem.

---

[51] Rationalizing is a functorial canonical construction that can be done to any simply connected space; the rationalized space has as homotopy (homology) that of the original space tensored with $\mathbb{Q}$.

[52] We have tacitly used the Hodge theorem as well to know how the Euler class of this bundle behaves on cohomology.

We can assume G_w is superperfect. (Note that G_w is trivial iff the element w is trivial in it.)

Let Ω be the fundamental group of our homology sphere with k generators. We can modify Ω by taking k amalgamated free products with G_w, always along w. Then, we take the $\mathbb{Z}^k$ central extension of this group. Now the proof is the same as that of the previous theorem: we produce a homology sphere for each element w of G, which, if w is nontrivial has nontrivial simplicial norm. If w is trivial, the fundamental group is trivial, and the homology sphere is the sphere. All of this is algorithmic, so this translates into a Turing machine, and completes our proof.

NOTES

The standard reference for the h-cobordism theorem is

J. Milnor. *Lectures on the h-Cobordism Theorem.* Notes by L. Siebenmann and J. Sondow. Princeton University Press, Princeton, N.J., 1965.

For its nonsimply connected extension (due to Barden, Mazur, and Stallings) and other applications:

J. Milnor. *Whitehead torsion.* Bull. Amer. Math. Soc. 72 (1966), 358–426.

C. Rourke and B. Sanderson. *Introduction to Piecewise-Linear Topology.* Reprint, Springer Study Edition. Springer-Verlag, Berlin, 1982.

J. F. P. Hudson. *Piecewise Linear Topology.* University of Chicago Lecture Notes prepared with the assistance of J. L. Shaneson and J. Lees. W. A. Benjamin, New York, 1969.

The latter two books contain such fundamental tools as transversality and general position in the PL category. These are also available in the smooth category. Useful introductory references are

V. Guillemin and A. Pollak. *Differential Topology.* Prentice-Hall, Englewood Cliffs, N.J., 1974.

M. Hirsch. *Differential Topology.* Corrected reprint of the 1976 original. Graduate Texts in Mathematics 33. Springer-Verlag, New York, 1994.

The other results discussed in section 1.1 are taken from Kervaire's paper. (Some of these results were obtained also by the Hsiang brothers.)

M. Kervaire. *Smooth homology spheres and their fundamental groups.* Trans. Amer. Math. Soc. 144 (1969), 67–72.

That Kervaire's theorem characterization of fundamental groups of homology spheres does not extend to dimension four was first observed in

J. Cl. Hausmann and S. Weinberger. *Characteristiques d'Euler et groupes fondamentaux des variétés de dimension 4.* Comment. Math. Helv. 60 (1985), no. 1, 139–144.

The extension of the theorem to other manifolds appears in a paper of J. Cl. Hausmann mentioned below.

That one cannot algorithmically recognize any manifold (as opposed to algorithmically decide whether or not two manifolds are diffeomorphic) is due to S. P. Novikov. It was proven in

S. P. Novikov. Appendix to I. A. Volodin, V. E. Kuznecov, and A. T. Fomenko. *The problem of the algorithmic discrimination of the standard three-dimensional sphere* (Russian). Uspehi Mat. Nauk 29 (1974), no. 5(179), 71–168.

The proof we sketched uses Grushko's theorem of combinatorial group theory. This theorem implies that $\gamma(G) = $ # of elements in the smallest generating set is additive for free products: $\gamma(G * H) = \gamma(G) + \gamma(H)$, and that therefore no group can have itself as a nontrivial free summand. Stallings has shown that Grushko's theorem (and some other close relatives including his own celebrated characterization of groups with infinitely many ends) is (are) intimately related to basic theorems about the topology of three-dimensional manifolds.

J. Stallings. *Group Theory and Three-Dimensional Manifolds.* A James K. Whittemore Lecture in Mathematics given at Yale University, 1969. Yale Mathematical Monographs 4. Yale University Press, New Haven, Conn., 1971.

The result about knots and its proof, as mentioned in the notes to the Introduction, is taken from

A. Nabutovsky and S. Weinberger. *Algorithmic unsolvability of the triviality problem for multidimensional knots.* Comment. Math. Helv. 71 (1996), no. 3, 426–434.

The negative results about hypersurfaces are proven in an appendix in this chapter. The positive result (existence of an algorithm to decide whether or not two embeddings of a simply connected manifold in another in codimension other than two are equivalent by an isotopy or by a homeomorphism) was proven in

A. Nabutovsky and S. Weinberger. *Algorithmic aspects of homeomorphism problems.* Tel Aviv Topology Conference: Rothenberg Festschrift (1998), 245–250. Contempory Mathematics 231. American Mathematical Society, Providence, R.I., 1999.

On a number of occasions one devises geometric algorithms by describing a smooth operation, then approximating by real algebraic varieties, and finally applying the Tarski-Seidenberg theorem. None of these steps is straightforward. For the most part, the first step will be what is concentrated on throughout these notes.

The second part is essentially the content of Nash's theorem (and subsequent refinements). Nash's theorem is found in

J. Nash. *Real algebraic manifolds.* Ann. Math. (2) 56 (1952), 405–421.

It asserts that every compact smooth manifold is a component of a real algebraic variety. In fact, they are all smooth varieties.

A. Tognoli. *Su una congettura di Nash.* Ann. Scuola Norm. Sup. Pisa (3) 27 (1973), 167–185.

Nowadays, many more polyhedra are known to be (PL-homeomorphic to) real algebraic varieties; see

S. Akbulut and H. King. *Topology of Real Algebraic Sets.* Mathematical Sciences
    Research Institute Publications 25. Springer-Verlag, New York, 1992.

It is worth noting that there are topological 4-manifolds that are not homeomorphic to
real algebraic sets (Freedman's E_8-manifold has that property because of work of Casson
or of Taubes). On the other hand, at least some of these do become homeomorphic to
real algebraic sets after crossing with a circle.

The basics of real algebraic geometry, including the Tarski-Seidenberg theorem and
algorithms that display the geometry of varieties can be found in

R. Benedetti and J. J. Risler. *Real Algebraic and Semi-Algebraic Sets.* Hermann, Paris,
    1990.

J. Bochnak, M. Coste, and M. F. Roy. *Real Algebraic Geometry.* Translated from the
    1987 French original; revised by the authors. Ergebnisse der Mathematik und ihrer
    Grenzgebiete (3) 36. Springer-Verlag, Berlin, 1998.

As we noted earlier, the Tarski-Seidenberg has a number of different useful reformula-
tions. The most "logical" of them, says that the theory of real closed fields is complete
and admits elimination of quantifiers. (That simply means that one does not ever need
to describe anything as "the set of x such that there is a y such that for all z one can find
a w which makes it true that for all u there is a v so that $P(x, y, z, u, v, w)$"—such a
set can always be described as "the set of x's such that $Q(x)$ holds.")

Moreover, there is also the equally useful and fundamental version that asserts that the
projection of a high-dimensional semialgebraic set (the set of solutions to a system of
polynomials equations and inequalities) is a semialgebraic set. (The projection of the
variety $x^2 + y^2 = 1$ onto the x axis is not a variety; it is a semialgebraic set.)

Nabutovsky's thesis, the topic of section 1.3, was published as

A. Nabutovsky. *Non-recursive functions, knots with thick ropes and self-clenching
    "thick" hyperspheres.* Commun. Pure Appl. Math. 48 (1995), 381–428.

As mentioned in the text, we will be discussing at greater length offshoots and further
developments of the ideas of this paper, as well as some of the analytic points that we
skipped over in this first summary.

The classification of homology spheres and its generalizations can be found in the papers

J. Cl. Hausmann. *Groupes de spheres d'homologie entiere* (French). C. R. Acad. Sci.
    Paris Sér. A 278 (1974), 1397–1400.

_____. *Manifolds with a given homology and fundamental group.* Comment. Math.
    Helv. 53 (1978), no. 1, 113–134.

The second paper of Hausmann goes beyond the situation of homology spheres to general
manifolds, as you might guess from its title.

As explained in appendix 1, these results essentially depend on homology surgery theory.
Homology surgery was invented by Cappell and Shaneson in order to place the analysis

of knot theory into a suitable general context that would encompass general codimension-two embeddings. It has had numerous other applications since.

S. Cappell and J. Shaneson. *The codimension two placement problem and homology equivalent manifolds.* Ann. Math. 99 (1974), 277–348.

The key calculational result of the Hausmann-Vogel theorem is that $\Gamma_n(\mathbb{Z}\pi \to \mathbb{Z}\pi') = L_n(\pi')$ if $\pi \to \pi'$ is a surjection with a perfect kernel. This was proven in Hausmann's thesis by a clever geometric construction (surgery using homology spheres rather than spheres), and later given algebraic treatments.

J. Cl. Hausmann. *Homological surgery.* Ann. Math. (2) 104 (1976), no. 3, 573–584.

J. Smith. *Homology surgery theory and perfect groups.* Topology 16 (1977), no. 4, 461–463.

P. Vogel. *On the obstruction group in homology surgery.* Inst. Hautes Études Sci. Publ. Math. 55 (1982), 165–206.

The homotopy pullback construction used in the proof of the classification theorem is standard homotopy theory: one can turn an arbitrary map into a fibration, and then pull that back. Spanier's text is an excellent source for basic homotopy theory.

E. Spanier. *Algebraic Topology.* Corrected reprint of the 1966 original. Springer-Verlag, New York, 1994.

Le Dimet has applied the idea of the Hausmann-Vogel theorem (using a localization to parametrize Poincaré data) to the problem of classifying links up to concordance.

J. Y. le Dimet. *Cobordisme d'enlacements de disques.* Mém. Soc. Math. France (N.S.) no. 32 (1988).

The results about knots in a "stable range" mentioned in the opening paragraphs of section 2.4 can be found in the paper of Farber that builds on the earlier work of Levine and Kearton:

J. Levine. *An algebraic classification of some knots of codimension two.* Comment. Math. Helv. 45 (1970), 185–198.

C. Kearton. *An algebraic classification of certain simple even-dimensional knots.* Trans. Amer. Math. Soc. 276 (1983), no. 1, 1–53.

M. Farber. *Isotopy types of knots of codimension two.* Trans. Amer. Math. Soc. 261 (1980), no. 1, 185–209.

After corollary 1, we remarked that, because $B\pi^+$ is simply connected, the main obstacle to computing its Hurweicz homomorphism is understanding its homology, and further remarked that one cannot even decide algorithmically whether a group is superperfect. This latter remark is due to C. Gordon in unpublished work. The former remark is due to the algorithmic nature of simply connected homotopy theory. In principle, the paper of Ed Brown gives a good deal of this:

E. Brown. *Finite computability of Postnikov complexes.* Ann. Math. (2) 65 (1957), 1–20.

In practice, one can actually do rational calculations. The original source for this is Sullivan's paper:

D. Sullivan. *Infinitesimal computations in topology*. Inst. Hautes Études Sci. Publ. Math. 47 (1977), 269–331.

Other useful expositions are

P. Griffiths and J. Morgan. *Rational Homotopy Theory and Differential Forms*. Progress in Mathematics 16. Birkhäuser, Boston, Mass., 1981.

S. Halperin. *Lectures on minimal models*. Mém. Soc. Math. France (N.S.) (1983) no. 9–10. (also a forthcoming book).

Not as well known, but actually quite useful, is Dwyer's "Tame" theory, where one inverts only finitely many primes in any given dimension, but the number of these increases with the dimension.

W. Dwyer. *Tame homotopy theory*. Topology 18 (1979), no. 4, 321–338.

However, we shall not need any of this theory later. (On the other hand, the perpicacious reader should have no trouble finding applications of this more refined theory almost every time we use rational calculations.)

Surgery theory itself, the basic theory of the classification of high-dimensional manifolds, can be said to have started with the paper of Kervaire and Milnor, which gave a classification of differential structures on the sphere.

M. Kervaire and J. Milnor. *Groups of homotopy spheres: I.* Ann. Math. (2) 77 (1963), 504–537.

Somewhat later, W. Browder and S. P. Novikov (independently) developed simply connected surgery, and, for example, gave the necessary and sufficient condition for a simply connected closed manifold to be homotopy equivalent to infinitely many other manifolds. D. Sullivan introduced F/Cat and used it to organize the classification into the surgery exact sequence (for the simply connected case). The nonsimply connected version is due to C.T.C. Wall. These developments are exposed in the books

W. Browder. *Surgery on Simply-Connected Manifolds*. Ergebnisse der Mathematik und ihrer Grenzgebiete 65. Springer-Verlag, New York, 1972.

C. T. C. Wall. *Surgery on Compact Manifolds*, 2nd ed. Edited and with a foreword by A. A. Ranicki. Mathematical Surveys and Monographs 69. American Mathematical Society, Providence, R.I., 1999.

More modern versions and expositions (which in no way replace the original sources) are

A. Ranicki. *Algebraic L-Theory and Topological Manifolds*. Cambridge Tracts in Mathematics 102. Cambridge University Press, Cambridge, 1992.

S. Weinberger. *The Topological Classification of Stratified Spaces*. Chicago Lectures in Mathematics. University of Chicago Press, Chicago, Ill., 1994.

The Novikov conjecture, and its relatives the Borel conjecture and the Baum-Connes conjecture, are central to high-dimensional topology, operator theory, and have applications in spectral theory, differential geometry, and pure algebra. Three quite different surveys (which, jointly, still don't tell the complete story, even as currently understood!) are

S. Weinberger, *Aspects of the Novikov conjecture.* Pages 281–297 in *Geometric and Topological Invariants of Elliptic Operators.* Contempory Mathematics 105, American Mathematical Society, Providence, R.I., 1990.

S. Ferry, A. Ranicki, and J. Rosenberg. *Novikov Conjectures, Index Theorems and Rigidity*, vols. 1 and 2, including papers from the conference held at the Mathematisches Forschungsinstitut Oberwolfach (Oberwolfach, Germany, 1993). London Mathematical Society Lecture Note Series 226 and 227. Cambridge University Press, Cambridge, 1995.

M. Gromov. *Positive curvature, macroscopic dimension, spectral gaps and higher signatures.* Functional Analysis on the Eve of the 21st Century, (New Brunswick, N.J., 1993), vol. II, 1–213. Progr. Math. 132.

The results mentioned regarding the Borel conjecture for discrete subgroups of $GL_n(\mathbb{R})$ are due to

F. T. Farrell and L. Jones. *Rigidity for aspherical manifolds with π_1 in $GL_m(\mathbb{R})$.* Asian J. Math. 2 (1998), no. 2, 215–262.

Those regarding the Novikov conjecture for all linear groups are due to E. Guentner, N. Higson, and me (currently available at http://www.math.hawaii.edu/~erik/research.html).

The Mayer-Victoris sequence for L-groups of amalgamated free products, so crucial for us, is due to Cappell.

S. Cappell, *On homotopy invariance of higher signatures*, Inven. Math. 33 (1976) 171–179.

The connection between the Novikov conjecture and nonexistence of metrics of positive scalar curvature was forged by J. Rosenberg:

J. Rosenberg. *C*-algebras, positive scalar curvature, and the Novikov conjecture.* Inst. Hautes Études Sci. Publ. Math. 58 (1983), 197–212.

Certainly M. Gromov and B. Lawson had already made the suggestion that these problems seemed connected. Rosenberg and S. Stolz have written a book on the current state of the art of identifying which manifolds have metrics of positive scalar curvature.

The simplicial norm was introduced by Gromov in his seminal paper

M. Gromov. *Volume and bounded cohomology.* Publ. Math. d'IHES. 56 (1982), 5–99.

It is based on earlier ideas of Milnor, Sullivan, and Thurston. It has been useful in many investigations where manifolds or metrics of nonpositive curvature are morally (or conjecturally) present. Because of the existence of doppelgangers that have "big pieces" with negative curvature, it becomes relevant to our investigations as well. In chapter 4, we will apply the homology spheres constructed by theorem 2 (of section 2.6) to get a priori inequalities on volumes in certain regions of moduli space. This is based on the work of Gromov, which we will explain there, that shows that the simplicial norm of a manifold bounds its volume from below, under suitable conditions on curvature.

Chapter Three

The Roles of Entropy

The goal of this short chapter is to begin the process of making the "logical method" more quantitative. We shall begin exploring the question of "how unsolvable" a problem must be for us to be able to use it for obtaining local minima, and we shall also examine the question of how many minima one is entitled to get. In the process of doing this, we will shift away from proofs by contradiction and toward more positive lines of thought.

Rather than deal immediately with spaces of Riemannian metrics with various properties, this chapter will deal with the energy functional on free loopspaces, that is, the problem of finding closed geodesics. The relevant variational theory for this problem is routinely taught in first-year differential geometry courses (we will review some of this in section 3.1), so that we can use it effectively as a toy version for the matters that will concern us in the final part.

Entropy has many meanings in the literature; it is hard to list them all. In all cases, it is a measure of the amount of complexity or information one has or a ratio of such, when one changes scale. In this chapter, no specific knowledge of any particular type of entropy is required: rather, we will be applying different versions of the idea of entropy.

3.1 THE PROBLEM OF CLOSED GEODESICS

Let M be a Riemannian manifold, that is, let (g_{ij}) be a smoothly varying family of positive definite quadratic forms on each tangent space $T M_m$. Using the metric, we can define $L(\gamma)$, the length of the curve g for $g : [a, b] \rightarrow M$ a piecewise smooth curve, by

$$L(\gamma) = \int_a^b \langle d\gamma/dt, d\gamma/dt \rangle^{1/2} dt,$$

where the inner product is given by the Riemannian metric (g_{ij}).

It is sometimes useful to use the energy of a curve in place of its length:

$$E(\gamma) = \int_a^b \langle d\gamma/dt, d\gamma/dt \rangle dt.$$

While the length is independent of reparametrization, the energy is not. One can always minimize the energy over all parametrizations by reparametrizing by arclength, that is, finding a parametrization $\gamma : [0, L] \rightarrow M$ where $L(g \mid_{[0,s]}) = s$. Consequently, while we will usually talk about minimizing length, really the arguments and formulas work better if one talks about energy.

Given any Riemannian metric on (a connected manifold) M, one gets an associated distance between points on M given by

$$d(p, q) = \inf \{L(\gamma) \text{ as } g \text{ runs over curves } g : [a, b] \rightarrow M$$

$$\text{where } g(a) = p \text{ and } g(b) = q\}.$$

This is actually a metric, and defines the original topology on M that underlies the discussion till this point.

A *geodesic* is a path that locally minimizes energy (and therefore arclength), that is, for sufficiently close points t, t', $\gamma(t)$ and $\gamma(t')$ are connected by no curve of smaller energy than $\gamma \mid_{[t,t']}$ (and, in particular, by no shorter curve). An equivalent definition is that geodesics are the critical points for the energy functional on the space of curves from $[a, b] \rightarrow M$ where one fixes the endpoints.

At any point m and any direction v in TM_m there is a unique geodesic emanating from m for which $g'(0) = v$ as follows from the Euler-Lagrange equations (this is the equation that $\delta E(\gamma) = 0$ for any infinitesimal variation of a geodesic γ, which can be rewritten as a second-order differential equation) and the basic existence and uniqueness theorems for differential equations. The famous theorem of Hopf and Rinow tells us exactly when geodesics can be continued for infinite time:

Hopf-Rinow Theorem: *Geodesics can be continued indefinitely iff (M, d) is complete as a metric space; either of these equivalent conditions implies that any two points p, q can be connected by a geodesic γ such that $L(\gamma) = d(p, q)$.*

In particular, if M is compact, then all geodesics can be continued infinitely, they are C^∞, and any two points can be connected by a length-minimizing geodesic.

Geodesics define the geometry of a manifold. The (sectional) *curvatures* of a manifold are *positive* at a point iff small geodesic "triangles" (i.e., unions of three geodesic segments at their endpoints) have angle sums that exceed π, and is *negative* if these sums are less than π. (For most manifolds, these angle sums can vary above and below π, even at a point, and if one mentions it at all, one says that M has mixed curvature.)

Manifolds whose curvature is everywhere positive or negative (or nonpositive or nonnegative or zero) are all extremely restricted and form interesting classes. This is a long and beautiful story that starts with the "second variational formula" connecting curvature with the second derivative (Hessian) of $E(\gamma)$ at a geodesic γ.[53] Let me start by sketching

Hadamard's Theorem: *Suppose M is a complete nonpositively curved manifold. Then in any homotopy class of paths connecting p to q, there is a unique geodesic.*

This implies, as we will see, that the universal cover of M is contractible.[54]

The second variational formula is the calculation of the Hessian of the energy functional at a critical point, that is, at a geodesic. The curvature enters in this formula and under the assumption of nonpositive curvature, it shows that the Hessian at any geodesic is positive definite, that is, that every geodesic is a local minimum. However, a function on a connected space (i.e., on a fixed component of the path space of curves connecting p to q) that has only (nondegenerate) local minima and no other critical points must have only a unique local minimum. (One can find another critical point by looking at $\min(\max(E(v))$ as v runs over curves in a path of curves connecting one local minimum to another, and one maximizes over all curves doing the connecting.)

Remark. Aside from the difficulties involved with dealing with an infinite-dimensional and noncompact space, this is a special case of the Morse inequalities.[55] The number of local minima is at least b_0 (the zeroth Betti number, i.e., the number of connected components) and b_0 is at least the number of local minima minus the number of critical points of index one. Assuming that there are none of the latter (as we have in our situation), we see that there can be only one local minimum in the component.

In fact, the way Morse theory works, one gets, up to homotopy type, a cell complex (CW complex structure) homotopy equivalent to the space on which the functional is defined, such that critical points of index i correspond to i-cells. Thus, if all critical points are local minima, that is, critical points of index zero, the components must be contractible. Letting $p = q$ and working with the component of the constant geodesic, one sees that ΩM, the loopspace so studied in homotopy theory, is contractible (homotopy equivalent to the complex that has one point in it!). Since

$$\pi_i(\Omega M) = \pi_{i+1}(M),$$

---

[53]This story continues with "comparison theorems" which we will discuss below and which form the context for much of our discussion in chapter 4.

[54]The usual argument for this is a little different; it shows that the exponential map at any point which sends TM_m to M, by sending v to the endpoint of the unique geodesic at m that starts in direction v, is a covering map.

[55]Interestingly, though, Morse initiated his theory by studying E on loopspaces to get information about their homology.

we discover that all the higher homotopy groups of M vanish, which is equivalent to saying that the universal cover of M is contractible.[56]

The argument we just gave actually proves the celebrated

Theorem   *(Lusternick and Fet) If M is a compact Riemannian manifold, then M possesses at least one nontrivial periodic geodesic (= closed geodesic).*

A periodic geodesic is one that satisfies $g(s + T) = g(s)$ for some s; it is called a closed geodesic because it represents a simple closed curve. (Note that not every geodesic that connects p to itself is closed: it could return "pointing in a different direction.") By nontrivial, we mean that it is nonconstant, that is, the particle moves.

Let us recap. If M is nonsimply connected, then one can minimize $L(\gamma)$ as γ runs over curves in a nontrivial homotopy class. If γ lies in the trivial homotopy class, then the minimum is realized by a constant curve, and one has to worry about that in the simply connected case. Consider now ΛM, the free loopspace of M, that is, Maps($S^1 \to M$). The critical points of E on ΛM are precisely the closed geodesics. We have an inclusion of M in ΛM as the constant loops. Evaluation at $t = 0$ gives a retraction back, $\Lambda M \to M$. In short, we have a fibration, $\Omega M \to \Lambda M \to M$, with a section.

Note that the minima of E (on the component of the trivial curves) are exactly where $E = 0$, that is, the constant curves. Suppose E has no other critical points, then by Morse theory ΛM would be homotopy equivalent to M with no additional cells glued on (the additional cells correspond, as before, to the new critical points). In other words, ΛM would deform retract to the constant curves. In yet other words, ΩM would be contractible, that is, the universal cover of M would be contractible. Since the nonsimply connected case was dealt with in the previous paragraph, we see that if M is simply connected and has no closed geodesics, it must be contractible, which is impossible.

Remark.   Lusternick and Fet's original proof did not use Morse theory, but went along similar lines. If M is simply connected then we can get an element α of $\pi_i(\Lambda M)$ along the above lines. Now one considers g realizing min max $L(\gamma)$ where one minimizes over maps $S^i \to \Lambda M$ homotopic to α (and maximizes length over the constituent curves making up the sphere). It is this curve that is Lusternick and Fet's closed geodesic.

Second (More Embarrassing) Remark

Of course, if a function has no critical points then it is Morse, so the above arguments are reasonably well founded. When there are some closed geodesics

---

[56]By the Whitehead theorem that says that a map between CW complexes is a homotopy equivalence iff it induces an isomorphism on homotopy groups, the inclusion of a point in the universal cover of M will satisfy this criterion.

and we are interested in understanding how many there are, the Morse theory techniques often apply only to "generic metrics," since the functional E, which depends on the metric g, could misbehave for a "small set" of g's.

The most pressing unsolved problem in this area is the following:

Problem: Does every compact Riemannian manifold possess infinitely many geometrically distinct closed geodesics?

"Geometrically distinct" means that one does not want to consider as different the result of starting a geodesic somewhere else along its trajectory ("rotating it") or going along it at a different speed (or several times).

There is a great deal known about this problem (see the notes) but it is not even known for metrics on the sphere in dimension > 2. A major source of the difficulty in this problem is identifying which cells, forced by Morse theory, are the results of going around a fixed geodesic many times.[57]

Let us now consider the question of existence of contractible (i.e., nullhomotopic) geodesics. These need not exist. In fact, as we saw above, on a closed nonpositively curved manifold (e.g., the torus with a curvature zero, i.e., flat) metric, there never are closed noncontractible geodesics.

These manifolds are all aspherical. The proof of the Lusternick-Fet theorem shows that whenever a manifold is not aspherical there are some closed contractible geodesics. In some sense, the aspherical case is, then, the set of manifolds for which the variational methods do not directly apply to produce closed contractible geodesics. In the next section we will prove the following:

Theorem   *If $\pi_1(M)$ has an unsolvable word problem, then M has infinitely many closed contractible geometrically distinct geodesics.*

We shall show in the appendix to that section that there are aspherical manifolds satisfying this condition. In particular, this provides us with a concrete example of a situation where a logical method gives rise to the existence of local minima.

3.2 ENTROPY OF FREE LOOPSPACES AND CLOSED CONTRACTIBLE GEODESICS

This section is devoted to proving and refining the theorem showing that an unsolvable word problem implies that there are infinitely many closed contractible geodesics.

---

[57]The rotation of geodesics is dealt with relatively straightforwardly by adjusting Morse theory to handle circle actions: the rotation of loops in ΛM.

Fake Proof. Suppose that the theorem were really false, that is, that there were no closed contractible geodesics. We will use this information to give an algorithm to solve the word problem for $\pi_1(M)$.

Given a word, represent it by a smooth closed curve in M. Now we will apply a curve-shortening process. We would like to say "the gradient flow of energy on ΛM," but this might not yet seem effective to the reader.

A concrete version of this is Birkhoff's curve-shortening process. Take our closed curve. Cover it by some small balls (at the scale of, say, half the convexity radius) centered at points on the curve. Now connect neighboring points by the geodesic between them. This will not be smooth (unless one, through, enormous luck got to a geodesic). But it will be shorter! Now connect the midpoints of these geodesics to each other by geodesics, and repeat the process.

Now, one of two things will happen to our curve: either it will keep on shrinking, until it lies within a small ball of M, or it will converge to a closed geodesic. If the first occurs, we know that the word represented the trivial element, and if the second, *on the assumption that there are no closed contractible geodesics*, we will be certain that the original word was nontrivial. Since there is no algorithm, one learns that there must be some closed contractible geodesic.

In fact, there must be infinitely many, because, otherwise, we could list them, and change the algorithm slightly: if we converge (to a rotated, multiple of) one of the geodesics on our list, we declare the word is trivial, and otherwise it is nontrivial.   End of Fake Proof.

There are a few criticisms one could level at this "proof." First, we have not discussed encoding of the Riemannian metric of M, and perhaps the theorem would fail for a "very transcendental ineffective" metric.

This is not a serious criticism if we reflect on the nature of the geodesics produced. Geodesics, generally, are merely critical points of energy, but the ones we produce are actually local minima. And not only are they local minima, but they are "deep local minima." (You might want to go back to the "Introduction and Overview" to review the definition of depth.)

Rather than get into technicalities let us consider an example.[58] Consider $y = x^3 - ax$ where a is a small positive constant. There is a local minimum at $x = (a/3)^{1/2}$. However, one can decrease the value of the function by going over a hump of size $2(1 - a/3)(a/3)^{1/2}$. It certainly is only a local minimum. As a result, a small perturbation, one of size $a(a/3)^{1/2}$, perturbs away this local minimum. On the other hand, no perturbation of the function of much smaller scale than that can remove this local minimum.

One can modify the curve-shortening process to produce deep local minima, by having the algorithm "sniff around" the neighborhood to try to converge to the deepest local minimum it can find (within a reasonable distance of the original curve). If this is not clear yet, draw a picture; it will be clearer in

---

[58] We shall discuss depths of local minima at much greater length in chapter 4.

just a page. The upshot, though, is that one can arrange to produce deep local minima.

Because of their depth, these local minima will persist for "nearby functionals," and in particular for the energy functional of a nearby metric. With a little care, one sees that any metric can be approximated by as nice and simple a metric as one likes, so that if the theorem (refined to assert that "deep index-zero" geodesics exist) holds for these, it holds in general.

However, there is another, more troubling difficulty with the alleged "algorithm." Suppose one has followed gradient flow for a very long time, and in the past millennium or two, the length of the curves did not shrink by more than a micron. Are we satisfied? Have we gotten near a closed geodesic? How can we tell? A real function can have an extremely small gradient in a neighborhood without the neighborhood being near any critical points.

As a result, we shall formulate an *analogous algorithm* in the PL setting based on a *nondeterministic gradient flow*. Besides being more transparent that this makes sense and works, this modification will allow us to connect our problem fairly clearly to Dehn functions.

Let us return to our manifold M. Let us triangulate it very finely (at a scale rather smaller than the convexity radius).[59] Any closed curve can be approximated by simplicial curves in this triangulation. The approximating curves are not unique, but any two sufficiently close curves are homotopic (by a straight line homotopy, using convexity), so they are homotopic to the original curve. Moreover, the approximation to γ can be chosen (with a fixed triangulation) to have length less than $cL(\gamma)$.

Now suppose that there are no closed contractible geodesics. Then the gradient flow for a contractible curve will ultimately shrink it to a small length. Let us consider the sequence of PL approximations to this flow. This will be a sequence of PL curves which will move from one to another in a "stop action" fashion. Nothing prevents us staying at some curve for a long time nor from returning to one having left it before (this, of course, cannot happen with a curve-shortening flow!). However, if we watch we will see the following:

Obvious Claim: *There will be a sequence of PL curves (in the fixed triangulation) whose lengths are less than $cL(\gamma)$ and which end up with a curve of length less than $c^{-1}L(\gamma)$.*

Note that if such a sequence exists, we can homotopically shrink γ by this sequence, that is, not necessarily by gradient flow, in an exponential in $L(\gamma)$ steps. Just eliminating any multiple appearances of the same PL curve will accomplish this, since there are only exponentially many such curves. (More

---

[59]It is a classical result of J. H. C. Whitehead that every compact Riemannian manifold has such an r, such that all r-balls are convex.

precisely, one shrinks g to a curve of a definite fraction of its length in exponentially many steps, and then repeats the process. This will give a sum of a linear-in-$L(\gamma)$ number of exponentials, so is exponential.)

Note that, as before, the argument works just as well if there are only finitely many geometrically distinct closed contractible geodesics.

We close this section with a sequence of remarks:

Remark 1. Although we have argued for closed contractible geodesics, the same argument works for producing infinitely many closed geodesics in any homotopy class that cannot be recognized.[60] In principle, one can do better by considering other homotopy classes, because the conjugacy problem can be much more complex than the triviality problem.

For instance, for groups with a solvable word problem but an unsolvable conjugacy problem, one can at least learn that there are infinitely many pairs of geometrically distinct freely homotopic geodesics. In every specific case I am aware of, one can prove that there are individual homotopy classes containing infinitely many closed geodesics.

Remark 2. The area of the annulus produced in the homotopy between nearby PL curves is approximately the length of these curves. Thus the sequence of curves produces a bounding disk of, at most, exponential area that γ bounds. This shows that all one needs is a superexponential Dehn function to guarantee infinitely many contractible geodesics.

Remark 3. Moreover, this argument is a much more affirmative restatement of the logical argument. It can be applied to specific curves. There is some constant K (depending on the Riemannian manifold M) such that if γ does not bound a disk of area at most $K^{L(\gamma)}$, then γ flows down to a nontrivial closed geodesic.

Remark 4. (Kevin Whyte) Three-dimensional Sol manifolds have exponential Dehn functions, but no closed contractible geodesics.

Remark 5. It is not hard to elaborate these methods as follows to deal with "depth." A superexponential Dehn function implies linear-depth local minima. (Just change the c in $cL(\gamma)$ that we used in searching for closed geodesics.) If the Dehn function exceeds a tower of exponentials, the depths will as well (just a shorter tower).

Remark 6. One can elaborate these methods to give examples where the number of closed geodesics of length $< L$ grows like c^L (this is the best possible; generically a Riemannian metric has at most exponentially many closed

---

[60] I do not know whether there is much to say about the variation in the recognizability of different conjugacy classes.

geodesics with length bounded by L): For a group with presentation, one can ask for the number of words of length $< L$ that represent the trivial element, but cannot be converted into another one by going through words of length CL, for some constant C. This is a coarse idea which gives one an estimate for the number of nullhomotopic geodesics in an arbitrary Riemannian manifold with fundamental group π. By considering the geodesics produced by curves of the form $[w, g]$ in $F * G$, where g lies in G and w runs over words in the nonabelian free group F that are, say, $1/4C$ times the length of g, the reader should be easily convinced that these different words really produce, after curve shortening, different closed geodesics. This kind of trick seems (for the present) special to the closed geodesic problem. When we move on to other functionals, while variants of the trick can be done, the methods using time-bounded Kolmogorov complexity that we are about to discuss seem to give sharper results.

Remark 7.   This argument was an entropy argument, in this case, a metric entropy argument. We used the fact that the number of balls of fixed radius needed to cover the space of curves of length less than L grows only exponentially in L in the course of developing our estimates on the Dehn function.

Appendix: Constructing Aspherical Manifolds by Reflection Groups

In this appendix, I will give a quick review of (a part of) Mike Davis's technique for constructing interesting aspherical manifolds.

Construction:

Let W be a manifold with a boundary. We suppose that ∂W is finely cellulated as a "manifold with corners." This means that every point has a neighborhood that looks like an open set in the boundary of an "octant"

$$\{(x_1, x_2, \ldots, x_n) \mid \text{all } x_i \geq 0 \text{ and at least one of them equals } 0\},$$

such that every "generalized face" is contractible (or even a cell). This can be achieved by taking the "Poincaré dual" to a fine triangulation. We produce an action of a group Γ on a manifold V, such that (1) $V / \Gamma = W$, and (2) Γ is a Coxeter group (see below) and is thus virtually torsion-free. We shall be interested in the manifold V / Γ' where Γ' is the torsion-free subgroup of finite index.

Coxeter groups are groups generated by reflections (abstractly involutions), so that the only relations are that $v^2 = 1$ for any element of the generating set, and $(vw)^{2m} = 1$ for some $m > 1$ (∞ allowed) for v and w in the generating set. The generators of Γ will be the faces of ∂W. We shall put in a nontrivial relation, that is, $(vw)^2 = 1$, whenever the faces corresponding to v and w meet.

Then V is simply $\Gamma \times W / \sim$, where the equivalence relation makes the face of W associated to v actually fixed by v. It is an interesting exercise to see that V is a manifold, smooth if W was. If W is aspherical, Davis shows that so is V. (Actually, Davis shows that if W is contractible so is V, but this implies the asserted variant.) A conclusion of all this is the following:

Proposition   *A group which is the fundamental group of a finite aspherical complex is a retract of the fundamental group of a compact aspherical manifold.*

We embed the finite complex in Euclidean space, and then take its regular neighborhood as W.

Corollary   *There are aspherical manifolds whose fundamental group has an unsolvable word problem.*

This is because there are groups with an unsolvable word problem π with a finite complex $B\pi$.

Remark.   One can use the Dehn function of a retract group to get a lower bound on the Dehn function of a group. It is easy to see that the group

$$\pi = \langle a, b, c \mid [a, c] = e,\ aba^{-1} = b^2,\ bcb^{-1} = c^2 \rangle$$

has a double-exponential Dehn function, and has $B\pi$ a 2-complex, so one can use it to produce aspherical 5-manifolds with infinitely many contractible closed geodesics. Using $F_2 * \pi$, as in remark 6 of section 3.2, one can produce an aspherical manifold where the number of closed contractible geodesics grows exponentially with length.

3.3 INTRODUCTION TO KOLMOGOROV COMPLEXITY

Kolmogorov complexity can be introduced in a number of ways and has many variants. Let us begin informally.

Suppose that S is a c.e. set, for which membership is algorithmically unsolvable. This means that I cannot determine on my own which integers in the interval $[1, n]$ lie in S. However, I can ask someone (I'd call this person a consultant, but the tradition is to use the name oracle) who does have a list of S.

Oracles will answer only yes/no questions, and they charge a lot per question, so it would be wise for me to ask something other than, is 1 in S? is 2 in S? and so forth (n questions).

Since S is c.e. I can be more clever and ask instead "How many elements of S are there in $[1,n]$?" (These are $\log_2 n$ yes/no questions.) Once I know the answer, I begin running the algorithm listing S and wait until that number of elements of $[1, n]$ appear, and then I can be content that I have discovered all the elements of S. It turns out that, in general, there is essentially nothing better to do.

Note that for some S I can do better (if I think of the right clever question). For example, S could consist of numbers of the form 2^t, where t lies in some other c.e. set T, and if I somehow know this, I can ask about $\log_2 \log_2 n$ questions.

Thus, we shall be interested in measures of the amount of information there is in a string of 0's and 1's, such as the characteristic function of a set. Information that I can get from doing a calculation we shall (temporarily) view as irrelevant. Thus, someone can send me the message "Compute the first $10^{10^{10}}$ digits of π," and it is succinct, even though it determines a very long string. (In terms of the concerns of an information theorist:) it is not hard to send all this information briefly (and reliably).

Definition. A describer is a computable partial function ϕ from the natural numbers onto $\mathbb{N}$. Let $K_\phi(x) = \log_2(\inf\{n \in \mathbb{N} \mid \phi(n) = x\})$.[61] K will be called the (ϕ-)complexity of x.

Definition. ϕ is said to be universal if, for any x, there is a c such that $K_\phi(x) < K_\psi(x) + c$.

In other words, a universal describer is as good as any other describer with maybe a one-time finite waste of a few bits. (As Robert Nozick pointed out, if you want your life to have meaning, just let it denote something like world peace, or universal love, and then it will mean that, and, in particular, have meaning![62] Obviously, any x can be described very briefly by a machine that is specifically designed to describe that x briefly, but that machine can be hard to describe and will likely describe other things less efficiently.)

It is an easy consequence of the existence of "universal Turing machines" that universal describers exist. Choosing one once and for all, we shall have an absolute notion of complexity of a finite string (or integer).

Theorem *Universal describers exist.*

Proof. Let $U(n, x)$ be a universal Turing machine, so that $U(n, x)$ computes (simulates?) the nth Turing machine on input x. Let $f(2^a(2b+1)) = U(a, b)$. We claim that f is a universal describer. For, let ψ be any other Turing machine, $\psi = U(n, -)$.[63] Now, clearly, $K_f(x) < K_\psi(x) + \log_2 n + 1$. QED

Henceforth, we shall assume that f is a universal describer and $K(x)$ will be short for $K_f(x)$.

Remark. For many purposes, there seems to be an analogy between $K : \mathbb{N} \to \mathbb{N}$ and the functionals that we will be considering. For instance, note that $\sup\{x \mid K(x) = n\}$ is closely related to $BB(n)$. This is true for the depth of the local minima produced by the logical method. If one thinks to search for $\sup\{x \mid K(x) = n\}$ by actually producing machines and inputs, that is quite

---

[61] Our $\log_2$ really means $[\log_2] + 1$, that is, the length of the binary string representing this integer.
[62] And who can forget the artist formally known as Prince (now denoted by a single symbol).
[63] Even if it is not a describer: then the ψ-complexity (x) for x not in the image of ψ is infinite.

similar to the idea of producing test curves (or the more complicated construction involving knots and hypersurfaces already discussed, or the connected sum metrics used in chapter 4, and the curves which are not nullhomotopic correspond to the pairs for which the machine does not halt—we never stop in our search for a nullhomotopy. The length of the nullhomotopy is (aside from considerations of the metric entropy of the moduli space we are discussing) the size of the integer output by the machine: after all, the length of the path from 1 to n is about n.

At times we have a string and are interested in estimating the complexity of computing the first n bits.

Definition. $KR_F(x) = \min\{\log_2(p) \mid F(p, i) = x_i \text{ for } i < \log_2(x)\}$.

And, as before, with this notion of complexity, universal F's exist.

Theorem *There is a c.e. set A such that, for all n, we have $KR(A|[1, n]) \geq \log_2 n$.*

This makes precise our earlier remarks about oracles. It is proved using a diagonal construction.

Proof. Define A by $x \in A$ iff $U(x, x) = 0$. So, if $U(x, x)$ converges to something other than 0 or does not converge at all, then $x \notin A$. A is clearly c.e. Suppose that $K(A|[1, n]) < \log_2 n$. Then there is some p, with $p < n$, such that $U(p, i) = \chi_A(i)$ for all $i < n$. Setting $i = p$ gives a contradiction. QED

For our purposes it is more convenient to work with "time-bounded Kolmogorov complexity." Heretofore we imagined ourselves able to compute arbitrary computable functions in manipulating the information given by the oracle, and tried to minimize the amount of additional information that had to be given to solve all instances of a decision problem up to a certain size. Now, one puts in a resource bound. (There are different possibilities such as space or time, but we will actually work with time.) We will not allow any of our machines to work for more centuries than there are bits of data to analyze.

This leads to the notion of $K^t(x)$ or $KR^t(A|[1, n])$ where t is a function of x or n, which represents the time bound. We shall not present details here, but shall merely note that this theory is more subtle and complicated than the theory without bounds. For instance, one does not even know whether the time-bounded complexity for accepting a string is the same as that of printing the string! (For us, $KR^t(A|[1, n])$ will be the string acceptance complexity.) Moreover, the "universality theorem" for the resource-bounded version is somewhat more complicated than that of the simple complexity discussed above.

Theorem *There is a universal partial recursive function ϕ and a constant c such that, for any ψ, we have $K_\phi^{ct \log t}(x) < K_\psi^t(x) + c$.*

In other words, in the formulation of the universal time-bounded complexity, in addition to the usual additive constant of ambiguity, one has to allow logarithmic factors of speeding up (or slowing down) for the machine time allowed.

The most useful feature of this strengthening of the notion of complexity is that we can no longer compute $A|[1, n]$ from the knowledge of how many elements it has: the necessary calculations can well exceed our time resource bound. As a result the following result becomes conceivable:

Theorem *There is a c.e. set A such that for any total computable function t, we have $K R^t(A|[1, n]) > c_t n$, where $0 < c_t < 1$ is a constant depending on t.*

The proof is not long—it is a little tricky diagonalization argument, where one is sure to list each Turing machine a positive proportion of the time, so that it gets defeated often—but we shall not give it. For our purposes, it is sufficient to know the statement and also that if one gives a specific time-bound monotonic total function t, then one needs only functions that compute for a time which are only exponentially larger to create a sequence that defeats all functions with time bound t. In the notes at the end of this chapter, we shall give a longer discussion of Kolmogorov complexity and its uses, and also some references.

3.4 COMPLEXITY AND CLOSED GEODESICS

Theorem *If π is a group whose time-bounded Kolmogorov complexity grows exponentially with word length, then any Riemannian manifold M with fundamental group π has an exponential number of contractible closed geodesics.*

Note that, in a finitely presented group, the number of words of length less than L grows exponentially in L. Thus, when discussing Kolmogorov complexity here, the natural unit is exponential, not linear.

The proof is quite similar to that of section 3.2. However, we shall use this opportunity to rephrase it slightly. Rather than discuss closed geodesics, we discuss the sublevel sets $E^{-1}([0, L])$.[64] These are rather unstable, and it is easier to deal with the image of $E^{-1}([0, L])$ in $E^{-1}([0, f(L)])$ for some function f. (In our previous proof f was just linear, but nothing prevents us from choosing f even larger, as long as it is computable.)

Remark. Usually homotopy theorists study the spaces $E^{-1}([0, \infty))$. However, in fact a great deal more information can be found in the sequence of sublevel sets. In our situation, the sequence for one metric gets mapped into the sequence for a different one, with a linear factor. Thus, for example, the question of whether (for sufficiently large L)

$$\ker\left(H_*(E^{-1}[0, L]) \to H_*(E^{-1}[0, \infty))\right)$$

---

[64]We are considering energy E as a functional on the trivial (i.e., nullhomotopic loop) component of ΛM.

is the same as $\ker\big(H_*(E^{-1}[0, L]) \to H_*(E^{-1}[0, AL])\big)$, for some A, is independent of the metric. Also, it is not hard to formulate questions about rank $\big(\text{im}(H_*(E^{-1}[0, L]) \to H_*(E^{-1}[0, \infty)))\big)$ that are independent of the metric.

Note that a lower bound on the number of closed contractible geodesics is $H(E^{-1}[0, L])$; the only component of this that need not contribute a closed geodesic is the component that contains the constant loops. Note that the Dehn function is simply[65]

$$D(L) = \inf \big\{ x \mid \ker \big(H_0(E^{-1}[0, L]) \to H_0(E^{-1}[0, \infty)))\big)$$
$$= \ker \big(H_0(E^{-1}[0, L]) \to H_0(E^{-1}[0, x]))\big) \big\}.$$

This very quickly proves (the alleged goal of section 1.2) that unsolvability of the word problem implies the existence of infinitely many closed contractible geodesics.

Let us suppose that

$$\text{rank}(\text{im}(H_0(E^{-1}[0, L]) \to H_0(E^{-1}[0, f(L)))))$$

grows exponentially in L. In other words, for a sequence of L's, one has this rank $= r(L)$ growing subexponentially. Then we could record for these L's $r(L)$ words of length at most L which we could use (as oracle information) to compress the calculation of the characteristic function of the nullhomotopic words, which would take only $\exp(f(L))$ time. Since $Lr(L)$ is still subexponential, this contradicts the assertion about the fundamental group.

Remark. In the statement of the theorem, there is no need to consider arbitrary computable t if one is just interested in closed geodesics. While it looks likely that linear functions can be used, it is only obvious to me that an exponential function suffices. In any case, once one restricts the time resource bound, one can apply this theorem to some groups with solvable word problems.

Remark. When the Kolmogorov complexity argument applies, it gives a significant qualitative improvement on the version of the logical method given in chapter 2, section 2.3. Those arguments show that, for infinitely many L, the L-sublevel set is disconnected, and even remains so within a much larger sublevel set. The complexity argument here, with its growth of components with L, shows that for all sufficiently large L these sets are disconnected.

Remark. There are groups (rather different from the ones I sketched at the end of section 3.2) for which the Kolmogorov complexity of the solution to the word problem grows exponentially. This follows from the technology of

---

[65] Here E can be thought of as a functional on the whole loopspace; this kernel cuts out the remaining components, but equally it can be the trivial component, in which case we are just talking about reduced homology. I chose this notation to maintain consistency with the theme suggested in the previous remark.

producing groups with nonsolvable word problems. For any c.e. set A one can produce a group GA such that there is a (computable) map $\mathbb{N} \to GA$ which sends n to a word of length around $c \log(n)$, such that n lies in A iff it represents the trivial element. A quick way to see this is to consider

$$HA = \langle x, y, z, a, b, c \mid w(x, y)zw(x, y)^{-1} = w(a, b)cw(a, b)^{-1} \rangle$$

as w runs through a set of words. List the words in the free group generated by x and y in some word length order, so the map $\mathbb{N} \to F\langle x, y \rangle$ sends n to a word of length proportional to $\log(n)$. Then send n in A to $w_n(x, y)zw_n(x, y)^{-1}w_n(a, b)^{-1}cw_n(a, b)$ in $F\langle x, y, z, a, b, c \rangle$ to define H_A. Then Higman embed (since the group is finitely generated, there can be only a linear length increase under the embedding).

The c.e. set with linear time-bounded Kolmogorov complexity then translates into a finitely presented group with exponential time-bounded Kolmogorov complexity.

NOTES

The elements of differential geometry are explained in many books. Some favorites, given our needs in this chapter and the next, are

I. Chavel. *Riemannian geometry: A Modern Introduction.* Cambridge University Press, Cambridge, 1993.

J. Cheeger and D. Ebin. *Comparison Theorems in Riemannian Geometry.* North-Holland, Amsterdam, 1975.

S. Gallot, D. Hulin, and J. Lafontaine. *Riemannian Geometry.* Springer-Verlag, Berlin, 1987.

P. Petersen. *Riemannian Geometry.* Springer-Verlag, New York, 1998.

Milnor's book on Morse theory contains a rapid introduction to Riemannian geometry and an extensive discussion of the application of Morse theory to loopspaces. In the next chapter, we will occasionally depend on deep ideas and results of Gromov, which are explained in the dazzling

M. Gromov. *Metric Structures on Riemannian and Non-Riemannian Spaces.* Progress in Mathematics 152. Birkhäuser, Boston, Mass., 1999.

As mentioned in the text, Morse introduced Morse theory in the infinite-dimensional setting of loopspaces. His original application was to the homology of ΩS^n. (He then inverted the processes and deduced that for any metric on the sphere there are infinitely many geodesics connecting any pair of points.) It was soon matched by Serre, using his well-known spectral sequence. Milnor's book contains some of Bott's spectacular applications of Morse theory to Lie groups, which are at the basis of index theory and K-theory (with its many connections to differential geometry, geometric topology, and mathematical physics—not to mention homotopy theory). Needless to say, there have been many applications of finite-dimensional Morse theory as well, in areas as diverse as symplectic geometry and combinatorics. It would be an extremely useful (if daunting) task to produce a modern successor to Milnor's book including some of the other applications.

The application to closed geodesics is difficult for a few reasons. First, one should use an equivariant version, because of the rotational symmetry in the problem. (Morse critical points are always isolated, and the critical points of E never are, because of this symmetry.) Second, it is not the case that E is Morse for every metric: it only is for generic metrics. (A very simple example, not particularly typical, is the round sphere where the primitive geodesics—up to rotation—form a Grassmanian.) Third, there is the problem of understanding multiple geodesics. Here one tries to show (at least) how the Morse index (which is the dimension of the negative definite subspace of the Hessian, i.e., the number of negative directions leaving the critical point) grows as one goes around the closed geodesic many times.

Despite all these problems, the following theorem asserts that one of the main qualitative conclusions of Morse theory holds anyway:

Theorem   (Gromoll-Meyer) *If the Betti numbers of ΛM are not bounded, then for every Riemannian metric on M, there are infinitely many closed geodesics.*

D. Gromoll and W. Meyer. *Periodic geodesics on compact Riemannian manifolds.* J. Diff. Geom. 3 (1969), 493–510.

Vigue-Poirrier and Sullivan showed that for simply connected manifolds whose rational cohomology algebra is not generated by only a single element (like the sphere or complex projective space), the hypothesis of this theorem holds.

M. Vigue-Poirrier and D. Sullivan. *The homology theory of the closed geodesic problem.* J. Diff. Geom. 11 (1976), 493–510.

The missing cases are handled by a combination of methods for generic metrics in

H.-B. Rademacher. *On the average indices of closed geodesics.* J. Diff. Geom. 29 (1989), 65–83.

_____. *On a generic property of geodesic flows.* Math. Ann. 298 (1994), 101–116.

For generic metrics on simply connected M, Gromov has related the rate of growth of the Betti numbers of ΛM with the number of closed geodesics (with length $< L$). (This is based on the idea, mentioned in section 3.4, of considering the sequence of homology of sublevel sets—in tandem with Sullivan's rational homotopy theory, mentioned in the notes to chapter 2.)

M. Gromov. *Homotopical effects of dilatation.* J. Diff. Geom. 13 (1978), 303–310.

It is irresistible to digress for a moment to say a word about other ideas involved in the study of geodesics. The paper of Rademacher makes use of the dynamical properties of geodesic flow. Note that by considering the geodesic that points in the direction of a tangent vector one gets a flow on the tangent bundle of any Riemannian manifold. This flow preserves the natural symplectic structure, and therefore one can apply the techniques of Hamiltonian dynamics to this flow. In particular, one studies the Poincaré return map associated with any periodic point, that is, closed geodesic; if the eigenvalues of this are sufficiently "good," then one can do a lot better in analyzing nearby periodic points and/or the Morse index of the iterated geodesics.

Another method that is based on the dynamical perspective is that of entropy (this is a third usage of the word "entropy" to appear in this chapter, but it is—at the level of the general idea—related to the other uses). This method gives better results for the problem of determining how many geodesic segments there are connecting two points than it does for the closed geodesic problem.[66]

Flows have (measure theoretic and topological) entropy. These measure how much accuracy one needs to have to predict within ε where a point will be after T seconds of flowing. Alternatively, they measure how spread out the image of an ε ball will be after T seconds. When the entropy is positive, there is "sensitive dependence on initial conditions."

One simple condition for positive entropy is that the number of elements of the fundamental group of length $< L$ (with the word length metric) grows exponentially with L. The reason for this is that the exponential map restricted to the unit ball of the tangent space at any point must cover at least a ball of that volume in the universal cover within L seconds. (Since geodesic flows on certain manifolds with exponentially growing fundamental groups arise as the Hamiltonian flows on phase spaces of simple physical systems, this gives a simple approach to "deterministic mechanical chaos.")

In general, there is a formula of Mane, which shows that the average number of length-L geodesics between pairs of points is quite closely related to the entropy. A useful review of this work can be found in

G. Paternain. *Geodesic Flows.* Progress in Mathematics 180. Birkhäuser, Boston, 1999.

Of course the most important open problem in this area is whether all Riemannian manifolds have infinitely many closed contractible geodesics.

A second problem that this work suggests is whether or not all finitely presented groups that are not virtually nilpotent have exponential volume growth. M. Gromov showed that they necessarily have superpolynomial growth, but A. Grigorchuk has given examples of finitely generated groups of superpolynomial subexponential growth.

M. Gromov. *Groups of polynomial growth and expanding maps.* Publ. Math. d'IHES 53 (1981), 53–73.

(This paper introduced the Gromov-Hausdorff convergence of metric spaces that plays such an important role in modern geometry, and in the next chapter.)

These entropic and dynamical methods give no information about the distribution of closed geodesics among the various homotopy classes. (Indeed, for negatively curved manifolds, the entropy is positive, but there are only unique closed geodesics in every free homotopy class.) The variational methods can take this into account, by fixing attention on each free homotopy class separately. I am not aware of any work that actually studies this.

---

[66]This is also an easier problem from the Morse theoretic point of view, as the discerning reader might have already noticed.

In the case of aspherical manifolds, these latter moduli spaces have the homotopy types of $B\mathbb{Z}(\gamma)$, where $\mathbb{Z}(\gamma)$ is the centralizer of g in π. (One should mod out by $\langle\gamma\rangle$ to deal with the rotational issue, and restrict to primitive—i.e. indivisible—elements to avoiding dealing with multiple geodesics.) These can have radically different homotopy types from one another, and can often guarantee the existence of many closed non-nullhomotopic geodesics. For instance, for a torus, it guarantees several closed geodesics in each nontrivial primitive homotopy class. It would be interesting to know how sharp these estimates are.

The logical method and its elaborations also work for producing closed geodesics in arbitrary free homotopy classes, as the reader can easily check. The theorem stated at the end of section 3.1 was first asserted by Gromov:

M. Gromov. *Hyperbolic manifolds, groups, and actions.* In *Riemann Surfaces and Related Topics.* Proceedings of the 1978 Stony Brook Conference, 183–213. Annals of Mathematical Study 97. Princeton University Press, 1983.

although I presume that a number of other people also observed this result.[67] The result weakening the bound from an unsolvable word problem to a superexponential Dehn function is due to me (I believe!), but heretofore unpublished.

Nabutovsky was the first to systematically exploit this idea as a principle for the study of variational problems, not just as a trick for getting information about closed geodesics. He pioneered the use of unsolvability results that do not seem to be part of the hypothesis (i.e., let M be a manifold whose fundamental group has an unsolvable word problem), for example, to get results that even apply to the sphere, like the results discussed in chapters 2 and 4. Another good example one can work out without much analysis, based on our current preparation, is the PL result from

A. Nabutovsky. *Geometry of the space of triangulations of a compact manifold.* Commun. Math. Phys. 181 (1996), 303–330,

which implies that there are (quantitatively many) minima of combinatorial functionals on the space of triangulations of a given manifold. We shall skip these results and leave them to remarks later.

To foreshadow one of the themes of the next chapter, it is worth thinking about what $\Lambda B\Gamma$ looks like near a nullhomotopic curve that does not bound any small-area disks. For quite a distance the free loopspace "thinks" that it corresponds to an essential curve. It might be relatively easy to see that g commutes with some subgroup H, which will produce a map of BH into the moduli space. Also, one might learn at different points that γ commutes with more and more elements, which will change what the neighborhoods look like as we move further out. Thus we can, in principle, find many different base points in the trivial component of $\Lambda B\Gamma$, around which enormously large neighborhoods look like different spaces. In order to study this, we will need dichotomy theorems for G; namely, that it is hard to tell whether or not γ is trivial or whether $\mathbb{Z}(\gamma) = K$ for some other group G. (Or one might have trichotomy or "multichotomy" theorems describing lots of different possible centralizers to get lots of different "local" structures.)

---

[67] I remember happily making this observation to Bob Brooks, who then showed me Gromov's paper. Bob taught me a lot over the years and I miss him.

In decisively giving examples where the logical method works and the straightforward variational one does not, we had recourse to Mike Davis's method of constructing aspherical manifolds. This method is one of the main ways of constructing aspherical manifolds, and was first applied by Davis to give examples of aspherical manifolds whose universal cover is not Euclidean space. However, it has much greater utility than that, as we have observed. For instance, Davis and Hausmann applied this construction to give the first nonsmoothable aspherical manifolds, and the first aspherical manifolds without a PL structure.

M. Davis. *Groups generated by reflections and aspherical manifolds not covered by Euclidean space*. Ann. Math. 117 (1983), 293–324.

M. Davis and J. Cl. Hausmann. *Aspherical manifolds without smooth or PL structure*. In *Algebraic Topology* (Arcata, Calif., 1986), 135–142. Lecture Notes in Mathematics 1370. Springer-Verlag, Berlin, 1989.

G. Mess also applied this construction to give aspherical manifolds whose fundamental groups are not residually finite.

G. Mess. *Examples of Poincaré duality groups*. Proc. Amer. Math. Soc. 110 (1990), no. 4, 1145–1146.

Our examples, of course, have this property as well. The only difference between our example and Mess's is the seed aspherical complex. The fact that there are finite aspherical (3-)complexes whose fundamental groups have unsolvable word problems follows readily from the monograph of Miller (see the references in chapter 1).

The reader can consult

M. Davis and T. Januskiewicz. *Hyperbolization of polyhedra*. J. Diff. Geom. 34 (1991), no. 2, 347–388,

M. Davis, T. Januskiewicz, and R. Scott. *Nonpositive curvature of blow-ups*. Selecta Math. (N.S.) 4 (1998), no. 4, 491–547,

R. Charney and M. Davis. *Strict hyperbolization*. Topology 34 (1995), no. 2, 329–350,

F. Paulin. *Constructions of hyperbolic groups via hyperbolizations of polyhedra*. In *Group Theory from a Geometrical Viewpoint* (Trieste, 1990), 313–372. World Scientific Publishing, River Edge, N.J., 1991

for other constructions of aspherical manifolds.

The other main collection of ideas relates to information, complexity, entropy, and the like. These are all interrelated concepts and have a long intricate history, especially when one looks back at imprecise historical approximations to the mathematically precise notions. The canonical references for these ideas are

A. Zvonkin and L. Levin. *The complexity of finite objects and the developments of concepts of information and randomness by means of the theory of algorithms*. Russian Math. Surveys 25 (1970), 83–124.

M. Li and P. Vitanyi. *An Introduction to Kolmogorov Complexity and Its Applications*, 2nd ed., Springer Graduate Texts in Computer Science. Springer-Verlag, New York, 1997.

The second reference is an especially friendly textbook source for study.

Kolmogorov complexity was actually first introduced in 1960 by R. Solomonoff in an attempt to deal with the problem of induction. He considered the phenomena in the "real world" as, ideally, being described by a binary string. The scientist tries to predict the next several bits from current knowledge.

The standard approach is to find a hypothesis and use it for prediction, and so on. However, there are infinitely many hypotheses compatible with a given finite mount of data. So Solomonoff tried an analogue of Occam's razor: try to choose the simplest (i.e., smallest program number) program that predicts this. He realized that picking once and for all gives a choice of the simplest that is almost independent of the universal Turing machine.

Of course, this minimum-complexity machine can be found only by trial and error. As with all scientific discovery, the "theory" can be invalidated by later experiments (later bits), or one might discover that there is a shorter program that has all the same convergents (at least empirically) and is, hence, to be preferred. But this hardly seems surprising to us.[68]

(Actually, Solomonoff's procedure is somewhat more complicated; he uses complexity ideas to define a universal a priori probability distribution, which he then combines with Bayesian ideas to get a probability distribution for the continuation of the series.)

Kolmogorov's interest came from the foundations of probability theory. He wanted a definition of "what is a random string." The idea is that the laws of probability are actually statements about random strings. It turns out that random strings satisfy all the laws of probability.

The approach to randomness is like this. A finite string is random if it has no description shorter than itself. (This version is dependent on the universal Turing machine used; however, if we ask for "significant compression" of the string it becomes more independent.) For any universal Turing machine, the fraction of n-strings that have complexity $\leq n - c$ is (at most) 2^{-c}. In other words at least 99% of strings can be compressed by no more than seven bits, and 99.9% by no more than ten. All of the computable laws of probability theory apply to random strings. (This last is work of Martin-Lof.)

This can be used to study a posteriori data. If we got a message that looked like 0001111110000001110001111111000000... we would note that there is some amount of regularity (tripling of bits) that surrounds a more random looking 01100101100. 00000000000 is less random than 01100010110 (obtained by flipping a coin), even though the former is just as likely an outcome from coin flipping as the latter. The reason is that strings like 0001111110000001110001111111000000 and 00000000000 can easily be compressed.

It might not surprise the reader that both Kolmogorov and Solomonoff give a lot of credit to Shannon—the inventor of information theory—for influencing their discoveries. And

---

[68] It seems that one of Kuhn's criticisms of older views on the philosophy of science would be included in a resource-bounded version, wherein the amount of time used in the course of implementing the program would be involved; after all, as Kuhn points out, in the practice of science, an earlier discovery will not be displaced merely by the existence of a simpler theory.

the connection between information and entropy goes back in embryonic form to Boltzmann, who recognized that the entropy of a gas is related to the number of microscopic configurations compatible with the macroscopic view. At a glance (macroscopically, shall we say?) one sees that 00000000000 is "all 0's" and has a unique microscopic possibility among 11-bit strings, and 000111111000000111000111111000000, being a tripled bit, has only 2^{11} possibilities versus the 2^{33} possibilities for a nontripled bit.

This compressibility also gives a more absolute definition of the amount of information there is in a string, as opposed to the idea in information theory, where one looks at strings only among their cohort of competitors (e.g., Morse codes of English sentences).

That one has a definition of a finite random string, or other finite random object, gives rise to a useful new method of proof: the "incompressibility method." In order to show that certain objects exist one can show that random objects have that property. This is like probabilistic proofs, which show that with probability one a property holds; these are often useful even in circumstances (perhaps especially in circumstances) where not even a single example can be exhibited.

One reason for this is the following proposition, strongly emphasized by G. Chaitin, the third independent discoverer of Kolmogorov complexity:

Proposition   *For any finite axiomatization of the integers, only finitely many integers can be proven random.*

Of course, as we mentioned, most numbers are quite random, but almost none of them can be exhibited!

The proof, which we leave to the reader, is based on the "Berry paradox" (just as Gödel's original proof was based on the "liar paradox"): "The smallest number that can be expressed in at most 10000 symbols" is exhibited in far fewer than 10000 symbols. The point is that it is impossible to tell within an axiom system how many symbols most numbers require.

This is, in turn, reflected in the algorithmic properties of $K(x)$ as a function of x (e.g., that it does go to infinity as x does, but it goes to infinity more slowly than any computable increasing function). We will see these things reflected in functionals on moduli space.

Another point, emphasized by Chaitin, is that there are sets that are low in the arithmetic hierarchy (they are of the form "for all m, there exists an n such that x has such and such relation with m and n") and are random, that is, incompressible. Chaitin has vigorously argued that this means that "mathematical truth is random" and that "God plays dice with mathematics."

Irrespective of one's views about the philosophical significance of this particular incompressibility result, the types of incompressibility possible somewhat lower on the arithmetic hierarchy do play an important mathematical role in our story. Chaitin's incompressibility is rather stronger than the incompressibility that c.e. sets possess (these are "there exists an n such that x has a certain relation with n"). That is Barzdin's theorem, and we discussed this in both its conventional form, where the first n bits can always be shrunk to $\log n$ bits, but according to Barzdin, not much more, and the time-bounded form, where it cannot be shrunk more than by a linear factor. Barzdin's theorem played an important role in Nabutovsky's theorem about exponentially large numbers of closed contractible geodesics on certain manifolds, and will also play an important role in the quantitative bounds on the number of critical points of Riemannian functionals to be proven in the next chapter.

Chapter Four

The Large-Scale Fractal Geometry of Riemannian Moduli Space

This part is the most difficult of this book. It contains basically one application of the methodology developed above, but it is an important one. An indirect application of the logical method together with some topological and/or group theoretic ideas teach us a great deal about how varied is the large-scale landscape of Riemannian moduli space (the space of isometry classes of metrics with bounded curvature on a manifold) for high-dimensional manifolds.

I believe that the picture we see here is typical of a wide range of moduli spaces that arise in Riemannian geometry, and, even more generally, for variational problems of nongeometric origin that have sufficient logical complexity.

Necessarily, the development we give here will be particular to our chosen moduli space. As a result we will need more material from differential geometry, surgery, psuedoisotopy theory, and, toward the end, transformation groups. The first section gives statements of some representative results, to orient the reader. I have not tried to give the "best results" in any direction: that seems premature. The theorems themselves are explained in sections 4.3, 4.4, and 4.7. The remaining sections provide the missing background, section 4.8 is devoted to what I think are some promising directions, and as usual we conclude with notes.

I point out that results about variational functionals appear relatively early in the chapter; the more difficult analyses are mainly necessary for discussions of geometry.

4.1 STATEMENT OF RESULTS

One of the great theorems of the nineteenth century was the uniformization theorem that asserts that every Riemannian metric on a surface is conformally equivalent to a constant curvature metric. After normalizing the curvature to be ± 1 or 0, the constant curvature metric is unique up to isometry. (This is not to say that there is a unique constant curvature metric on a surface—this is true only for S^2—rather that conformally equivalent constant curvature metrics are isometric.)

The upshot of this is that for the surface Σ_g of genus g, there is a deformation retraction

$$M_g \to \mathrm{Riem}(\Sigma_g)/\mathrm{Diff}(\Sigma_g),$$

where M_g is the moduli space of curves of genus g, in the language of the algebraic geometers, or simply the isometry classes of constant curvature metrics on the surface. This space is not quite compact, but it is a quasiprojective variety; it has been an object of intensive study for well over a century and it continues to fascinate.

Remark. $\mathrm{Riem}(M)/\mathrm{Diff}(M)$ is the quotient of the space of Riemannian metrics on M by the action of the diffeomorphism group of M (which pulls back metrics). It is, equivalently, the space of isometry classes of Riemannian metrics on M.

In dimension three, uniformization cannot be true in too naive a sense: no nontrivial connected sum of manifolds has a constant curvature metric. A more sophisticated example is the Heisenberg manifold $U_3(\mathbb{R})/U_3(\mathbb{Z})$, which is the quotient of the space of upper triangular 3×3 matrices with real entries modulo those with integer matrices. Thurston has completely analyzed the situation and found that there are eight geometries that can conjecturally account for all 3-manifolds.[69] (In addition to making this audacious conjecture, Thurston revolutionized three-dimensional topology by proving it in a great many cases.)

If one includes in this conjecture the idea that the "geometric metric" should be obtained by a variational technique, as is the case for surfaces, one would again get an analysis for (irreducible) 3-manifolds of $\mathrm{Riem}(M)/\mathrm{Diff}(M)$. For instance, if M is hyperbolic (which is, by far, the most common geometry), recent work of Gabai, combined with Mostow rigidity and results of Thurston, gives a topological proof that there is a homotopy equivalence

$$\mathrm{Riem}(M^3)/\mathrm{Diff}(M^3) \to \text{ a point } = \{\text{hyperbolic metrics on } M^3\}.$$

(For other geometric 3-manifolds, there are obstacles to producing such an equivalence. Even for the 3-sphere one does not yet know this. However, were "the elliptic case of geometricization" known, together with "the Smale conjecture for space forms," the same conclusion could be deduced for S^3.)

In higher dimensions, nothing like geometricization can be true. On the 4-sphere, there are already nonlinearizable differentiable symmetries; on the 6-spheres, there are diffeomorphisms not isotopic to isometries. We shall see below (in section 4.6) that symmetries contribute singular strata to Riem/Diff, and indirectly can complicate its topology.

Pseudoisotopy theory gives rise to some even more dramatic failures of our most naive and idealistic hopes. Despite this, the search for variational principles that could lead to useful special metrics goes on.

---

[69]As I revise this chapter, I can happily note that Perlman has circulated what could be a proof of this conjecture. See http:/front.math.ucdavis.edu/math.DG/0211159 and its successor.

The main goal of this chapter is to give information about the geometry of $\text{Riem}(M^n)/\text{Diff}(M^n)$ for $n > 4$.[70] As in the constant curvature case, we shall make the normalization that $|K| \le 1$. This space has been the object of many deep investigations; while we will review some of the main conclusions of the vast literature on the homotopy types of these spaces, our focus instead is on its geometry, and on phenomena not forced directly by the homotopy theory.

Definition. $\mathcal{R}(M) = \{$Isometry classes of Riemannian metrics on M with $|K| \le 1\}$.

The way we get information about $\mathcal{R}(M)$ will involve studying the local minima for variational problems, but rather indirectly.

Theorem 1 *If M is a compact smooth n-manifold with $n > 4$, the functional $D =$ diameter*

$$D : \mathcal{R}(M) \to \mathbb{R}$$

has infinitely many local minima. Indeed, for any computable function $f(D)$, there are infinitely many local minima that are $> f(D)$ deep.[71] The number of such local minima grows at least like c^{c^D} for some $c > 1$.

The same also holds for other functionals like $D^\alpha \text{Vol}$, for any $\alpha > 0$ (but with a single exponential lower bound on their number).

We have to admit that the local minima are not (as far as we know) smooth of class C^2. One only gets a somewhat weaker class of smoothness $C^{1,\alpha}$ for all $\alpha < 1$.

Recall that the depth of a local minimum is the amount that the value of the function must rise during a path from that point to a point where the value of the function is smaller. To say that a local minimum has depth $> f(D)$ means that the depth is at least $f(D)$, where, by abuse of notation, D is the value of the functional D at the local minimum.

It is worth noting that for $M = T^n$, the n-torus, there is no global minimum; $\mathbb{R}^n / \varepsilon \mathbb{Z}^n$ as $\varepsilon \to 0$ is a sequence of metrics which "falls off the space." For any M, $\mathcal{R}(M)$ is connected, so one is not entitled to more than one local minimum (the global minimum) by the usual methods.

In section 4.3 below we will indicate why there are infinitely many local minima for D on $\mathcal{R}(M)$ for at least some simply connected 4-manifolds. However, some of the more refined quantitative information is lacking even for these.

In theorem 1, the larger the function f, the stronger the conclusion: the deeper the local minima produced must be. However, very deep local minima must, by definition, be distributed rather sparsely about the space. Thus, there

---

[70] Many of the results apply to certain 4-manifolds. It is an interesting question, which we will occasionally pursue in remarks, to speculate on the extent to which our analysis holds for all 4-manifolds.

[71] More precisely, the maximum depth of all the local minima with $D(\text{minima}) < d$ grows like the busy beaver function.

is value in trying to find local minima that are not as deep. Theorem 2 has many variants, all of which say, in some sense, that there are infinitely many scales at which it can be truly asserted that the local minima that are deep at that scale are dense at the same scale. We shall use classes of functions to measure scales.

Definition. Let f and g be increasing functions. We say that f and g have the same growth rate if, for some c, we have $f(c^{-1}x) < g(x) < f(cx)$.

$\mathcal{R}(M)$ has many metrics on it. For us, there are two important ones (more details can be found in the next section). The first is the Gromov-Hausdorff metric. It is a general metric on the space of all compact metric spaces that measures how far apart the two metric spaces must be, no matter where they would be commonly situated. Different topological types are at finite distance in this metric.

The path distance on $\mathcal{R}(M)$ is defined as follows. Set ε as some fixed very small real number. (In the next section we will describe how to pick this ε; for now it is unimportant.) Given two points p, q in $\mathcal{R}(M)$, consider all chains that connect p to q, $p = x_0, x_1, \ldots, x_n = q$, such that the distance between any consecutive x's is $< \varepsilon$. Now let the length of this chain path be $\sum d_{GH}(x_j, x_{j+1})$. The path distance is (by definition) the infimum of the lengths of all connecting chains. The ε that we choose will be sufficiently small to ensure that (within our class of Riemannian manifolds) two points of finite path distance are necessarily homeomorphic.

Remark. Defining path metrics in this fashion is a convenient way of studying the geometry of spaces without dealing with the issue of local rectifiability.

Theorem 2 *Let M^n be a closed smooth n-manifold with nonzero Euler characteristic[72] and dimension $n > 4$, and f be an increasing stopping time of a Turing machine for a computable set, such that f grows at the same rate as $\exp(f)$. Then there are functions g and h in the same growth type as f, such that every point in $\mathcal{R}(M)$ is at most $g(D)$ away from a local minimum of depth $> h(D)$ in the path metric; the value of the function at this local minimum can be taken to be $< 2D$.*

Remark. A variant of theorem 2 that is true for arbitrary M is somewhat less explicit and goes like this. There is a computable function z such that if f (is a stopping time function for a Turing machine which) grows faster than iteration of z, that is, if $f \sim z(f)$, then the theorem holds for the function f. I believe that z can always be chosen to be exponential. It is also possible to show that there is always a "Gromov-Hausdorff dense" open set for which the exponential version holds.

It is easy to give infinitely many scales f to which theorem 2 applies. Let $d[n] = d^{d^{[n-1]}}$. That is, $d^{[n]}$ is a tower of n exponentials of d. Let $f_a(d) = d^{[d^a]}$.

---

[72]The simplicial norm or any other characteristic number will do.

If $a < b$, then f_a grows more slowly than f_b; if a is a real number whose decimal expansion can be computed sufficiently quickly, then f_a is a stopping time of such a machine.

Theorem 2 asserts that with this very wide definition of the "same scale," we can find local minima at the same scales of depth and of density. There are deeper and deeper local minima, but these are more and more sparsely distributed within the space. However, for the shallowest local minima, one does not have to go more than a tower of exponentials away to find one.

Remark. It is possible to give rather more refined statements than theorem 2, but we shall not pursue these here. (For one example, it is possible to *bound* the number of exponentials needed to get to a local minimum. See section 4.8 for a discussion of what Nabutovsky and I believe to be correct.)

In the introductory chapter, we phrased this result as the assertion that $\mathcal{R}(M)$ is a large-scale fractal. Now we shall go a bit further and apply these results to get additional information about the geometry of $\mathcal{R}(M)$. We shall be concerned with how the "homological filling geometry" differs around different points of $\mathcal{R}(M)$. The local minima describe for us convenient base points around which to examine $\mathcal{R}(M)$. (It is like exploring the world by touring around the major capitals. One does see various landscapes, and great differences between one place and another, but, of course, one never explores the great uninhabited regions.)

The following is typical of a general type of theorem that we will discuss below:

Theorem 3 *For every manifold M, every k, and every computable function f, there are infinitely many points tending to infinity such that $\mathrm{Im}\,(H_k(B(D(p), p)) \to H_k(B(f(D(p)), p)))$ is arbitrarily large. (This is true with arbitrary field coefficients.)*

The reason this is interesting is that, although $\mathcal{R}(M)$ might have very little homology, the theorem asserts that there are points where the space for arbitrarily long times looks as if it has lots of homology. (One can arrange for these extra cycles to die at particular "rates.")

A different type of geometric study can be made of filling functions.

Definition. F is an H_k-*filling function for* X if, choosing a base point p and any $r \gg 0$, we have $\mathrm{Im}\,(H_k(B(r, p)) \to H_k(B(F(r), p))) = \mathrm{Im}\,(H_k(B(r, p)) \to H_k(X))$. A similar definition can be made for the fundamental group or higher homotopy.

Definition. F is an *absolute* H_k-*filling function for* X if, for all p and any $r \gg 0$, we have $\mathrm{Im}\,(H_k(B(r, p)) \to H_k(B(F(r), p))) = \mathrm{Im}\,(H_k(B(r, p)) \to H_k(X))$. And, again, one can formulate a similar definition for the fundamental group or higher homotopy.

The following two theorems contrast these notions.

Theorem 4   *If M is a closed smooth n-manifold, n > 4, there are no computable absolute filling functions for $\mathcal{R}(M)$ for π_1.*

Theorem 5   *For $M = S^n$, $n > 4$, there is no computable filling function for $H_5(\;;\mathbb{Q})$; for products of Kummer surfaces[73] or hyperbolic manifolds, the filling functions are computable.*

Thus, in theorem 4 we see that it is always the case that there are faraway points around which cycles fill in at a very slow rate. However, the slowness of the rate is often comparable in some weak sense to its distance from the base point for the "good" manifolds of theorem 5. On the other hand, for the sphere, even around the round metric, the terrain is quite rugged. One can bound, for all computable f, the rank of Ker: $(\mathrm{Im}\,(H_k(B(r, p)) \to H_k(B(f(r), p))) \to H_k(\mathcal{R}(S^n)))$. It grows at least exponentially in r^{n-2}.

Theorem 4 is proved by examining the structure of $\mathcal{R}(M')$ when M' is a doppelganger of M, and using the logical method. Computable invariants of $\mathcal{R}(M')$ for doppelgangers must resemble the invariants one sees in open sets of $\mathcal{R}(M)$. Theorem 5 is based on a similar principle, but is based on an analysis of the complications forced by the singularities one finds in $\mathcal{R}(M)$.

Remark.   In the situation of 2- and 3-manifolds, there are other analogues of moduli space, namely, pointed moduli space and Teichmüller space, which for some purposes have advantages. It is not hard to define analogues of these spaces in the higher-dimensional situation to which similar methods to those applied in theorems 1–4 apply. Typically, these spaces "break symmetry," so the method applied in theorem 5 does not.

Definition. $\mathcal{P}R(M) = \{$Isometry classes of Riemannian metrics on M with $|K| \leq 1$ together with the choice of a base point p and a basis for $TM_p\}$.

$\mathcal{P}R(M)$ is the "pointed Riemannian moduli space." Since any isometry of a connected Riemannian manifold which fixes a point and induces the identity on the tangent space of that point is the identity, $\mathcal{P}R(M)$ has no singularities.

Definition. $\mathcal{T}(M) = \{$Isometry classes of "polarized" Riemannian metrics on M with $|K| \leq 1\}$; that is, they consist of triples (M, g, f), g a metric on M, and $f : M \to M$ a homotopy equivalence.

To see how this can be useful, suppose that M admits a negatively curved metric; for instance, one can "flow" the map f to the unique harmonic map homotopic to f from (M, g) to this negatively curved metric. In any case, $\mathcal{T}(M)$ is rather simpler to analyze than any of its cousins, and is closest to surgery and pseudoisotopy theories.

---

[73]The Kummer surface is a simply connected spin 4-manifold with signature one.

As mentioned above, despite these advantages, theorems 1–4 do apply to these spaces as well. (In other words, choices of point frames and of polarizations do not significantly flatten basins.)

4.2 NEOCLASSICAL COMPARISON GEOMETRY

In this section, I will review some modern ideas regarding "comparison geometry" with two-sided curvature bounds. I chose to call this "neoclassical" rather than modern, because in the modern period, one tends to work with weaker curvature bounds than those dealt with here.

The archetypical theorem in this subject, and the one most germane to our discussion, is the result of Cheeger's thesis, the finiteness theorem:

Theorem   *The set of n-manifolds with vol $> v > 0$, $|K| < C$, and diameter $< D$ contains only finitely many diffeomorphism classes.*

Here K denotes the sectional curvature, and we are assuming the curvature bound for all 2-planes in all of the tangent spaces at all points.

We will first sketch the proof of a rather weaker statement, that there are finitely many homotopy types.

Definition.   If X and Y are compact metric subspaces of a metric space Z, then the *Hausdorff distance* between X and Y is defined by $d_H(X, Y) = \inf\{r \mid X$ is contained in the r-neighborhood of Y and Y is contained in the r-neighborhood of $X\}$. If X and Y are abstract compact metric spaces, then the Gromov-Hausdorff distance between X and Y is defined by $d_{GH}(X, Y) = \inf d_H(X, Y)$, where the inf is taken over all metric spaces Z, containing X and Y isometrically. A moment's reflection shows that without loss of generality Z can be chosen to be $X \cup Y$ with some metric extending the given ones on X and Y.

We will not give the proof that d_{GH} actually defines a metric on the set of isometry classes of compact metric spaces. (The hardest check is that $d_{GH}(X, Y) = 0$ implies that X and Y are isometric: this is done using the proof of the Arzelà-Ascoli theorem.) The Gromov-Hausdorff metric makes it possible to talk about limits of metric spaces, even when the underlying topological spaces have nothing in common. It is also not hard to tell, at least in theory, when a sequence of metric spaces has a limit point:

Proposition   *Every subsequence of a sequence $X_1, \ldots, X_n, \ldots$ of compact metric spaces has a convergent subsequence iff they have uniformly bounded diameter,[74] and for every $\varepsilon > 0$, there is an $N(\varepsilon)$ such that all the X's can be covered by at most $N(\varepsilon)$ balls of radius ε.*

---

[74]For path-connected spaces, this condition follows from the second condition.

The idea is also that of Arzelà-Ascoli; one approximates the X's by the finite metric spaces, and takes a limit of metrics on bigger finite sets to produce the limit space.

Note that Cauchy sequences therefore converge: If $X_1, \ldots, X_n, \ldots$ is Cauchy, then an $N(\varepsilon/2)$ chosen for a very far out metric space (using its compactness) will work as an $N(\varepsilon)$ for all sufficiently far out spaces by the Cauchy condition. (The bounded diameter is even more obvious.)

Bounds on curvature (or even just Ricci curvature) suffice to give a bound for $N(\varepsilon)$. This follows from the Bishop-Gromov inequality. According to that inequality, for any manifold M with Ricci $\geq -(n-1)$ (e.g., if the ordinary sectional curvature is ≥ -1) and $s < r$, one has that

$$\operatorname{vol} B(r)/\operatorname{vol} B(s) \leq \operatorname{volh} B(r)/\operatorname{volh} B(s)$$

where volh is the volume of the corresponding ball in hyperbolic space. In particular, if M has diameter D, then there can be at most

$$\operatorname{volh} B(D)/\operatorname{volh} B(\varepsilon/2)$$

disjoint $\varepsilon/2$ balls in M. Therefore, this number suffices as an $N(\varepsilon)$. Let us summarize this:

Proposition *Any set of manifolds with a lower bound on (Ricci) curvature and an upper bound on diameter is precompact in Gromov-Hausdorff space.*

The upper bound on curvature and lower bound on volume in the finiteness theorem are applied via the following lemma:

Lemma *The bounds on curvature, volume, and diameter give rise to a lower bound on the injectivity radius.*

This lemma follows from the Rauch comparison theorem and the first variational formula. We shall not require its proof except to note that the bound on the injectivity radius has a quite explicit dependence on the various parameters.

We can combine the above remarks quite simply to get a proof of the finiteness of homotopy types.

Claim *If two n-manifolds M, N with injectivity radius $i > 0$ have $d_{GH}(M, N) < i/2^n$, then M and N are homotopy equivalent.*

Proof. Choose a very dense set of points in M and N that are vertices for very fine triangulations. We can match these sets with each other using the Gromov-Hausdorff closeness. Then, since we are always mapping well within the injectivity radius, we can constantly extend the maps to higher and higher skeleta. This builds continuous maps between M and N. All composites are close enough to the identity (no point ever gets moved very far) to be able to show that these composites are homotopic to the identity.

Using this claim, the finiteness of homotopy types follows.

Remark. It would be natural to try already to execute the logical method (assuming we had a lower bound on volume): We have "doppelganger" homology spheres Σ_k which are either nonsimply connected or are the sphere. We consider $M\#\Sigma_k$ and try to connect it in the relevant Gromov-Hausdorff space to M. If we could, then we would learn that Σ_k was a sphere and that $M\#\Sigma_k$ was actually diffeomorphic to M (and not just homotopy equivalent to it). The trouble is that, while we embedded our manifold part of Gromov-Hausdorff (GH) space into a compact space, which has finite covers that are very simple to construct at any scale, we need to know whether there is a path in GH space of *manifolds with the injectivity radius condition* able to turn the path into a homotopy equivalence. Thus, we need *effective precompactness*, not just an effective proof of precompactness, where all the ε's and $N(\varepsilon)$'s can be explicitly computed. That is, we need, for each ε, a choice of $N(\varepsilon)$ explicit manifolds which form the ε-net.

The method we use for getting effective precompactness is closely related to one for getting the diffeomorphism finiteness and information about the geometry of the limit spaces that arise. The basic idea is that we already know what the pieces of the manifolds we are considering are: They are $N(\varepsilon)$ balls (metric balls are smooth balls) glued together by diffeomorphisms. Thus, one can try to control the gluing maps.

Not surprisingly, there is a considerable literature about "best coordinate charts" and "smoothing Riemannian metrics" to make them better. By gaining extra smoothness, it becomes easier to make effective covers: this is the upshot of analytic theorems like that of Arzelà-Ascoli or the Reillich lemma. Similar things work topologically; we will content ourselves with summarizing one of the conclusions.

Smoothing Theorem

On the set of complete Riemannian manifolds with $|K| \leq 1$, there exists for all $e > 0$ a smoothing operator $g \mapsto S_\varepsilon g = g^*$ such that

1. $e^{-\varepsilon} g \leq g^* \leq e^\varepsilon g$;
2. $|\nabla_g - \nabla_{g^*}| < \varepsilon$;
3. $|\nabla^k R| < C(n, \varepsilon)$; and
4. The value of g^* at a point depends only on the values of g in a specified neighborhood of the point.

Furthermore, any isometry of g is also an isometry of g^*.

The values of the constants $C(n, \varepsilon)$ are computable from the proof, and we shall not discuss this. Once one has a degree of smoothness, with bounds, then it is easy to get explicit C^k approximating polynomials with bounds on degrees and coefficients.

The upshot is that associated with any smooth Riemannian manifold (with given bounds on curvature, volume, and diameter) we can associate a GH-close metric with semialgebraic charts, gluing maps, and metric, where one has bounds on the "complexity" of the situation: that is, the degrees, coefficient sizes, and gluing patterns involved. This set of metric semialgebraic manifolds is itself a semialgebraic set, and one can compute GH distance on it, and find an effective cover of it. Replacing the ε's by $\varepsilon/2$'s, one immediately gets an effective cover of the original Gromov-Hausdorff space.

Remark. For our applications to filling functions, and the like, in the later sections, note that because the smoothing theorem is proven using a flow produced by a geometric partial differential equation (PDE), one can (effectively) move chains and cycles from the moduli space into the semialgebraic approximations.

4.3 EXISTENCE OF EXTREMAL METRICS

In this section, we will complete our discussion of theorem 1. The method will be the usual: We will produce many connected components of $D^{-1}(0, d)$ (which remain disconnected from each other in $D^{-1}(0, d + f(d))$) with a lower bound on volume. This is accomplished by means of our earlier work in chapter 2 on producing hard to recognize homology spheres with nonzero simplicial norms together with an important lower bound (due to Gromov) on volume in terms of the simplicial norm.

Then we will see that there is a $C^{1,\alpha}$-metric on M at the bottom of each of these components (henceforth called "basins"). This step is a continuation of the ideas of the previous section and is a feature of the Gromov-Hausdorff convergence.

Finally, we will return to the issue of how many basins can actually be produced. This is done using a construction of Nabutovsky that directly produces c^{c^d} of them. We remind the reader of the statement we are trying to prove:

Theorem 1 *If M is a compact smooth n-manifold, $n > 4$, the functional $D = $ diameter*

$$D : \mathcal{R}(M) \to \mathbb{R}$$

has infinitely many local minima. Indeed, for any computable function $f(D)$, there are infinitely many local minima that are $> f(D)$ deep.[75] The number of such local minima grows at least like c^{c^d} for some $c > 1$.

---

[75]More precisely, the maximum depth of all the local minima with $D(\text{minima}) < d$ grows like the busy beaver function.

Note that as in our earlier work on closed geodesics, naively an argument by contradiction shows only that for infinitely many values of D there are $f(D)$-deep local minima, but the quantitative bound implies that this is true for all sufficiently large D. Although we will argue by contradiction, the argument can be rephrased along the lines of chapter 3, section 3.2 to be more positive and "construct" the local minima.

We shall show first that there are components of $D^{-1}(0, d)$ that remain disconnected from each other in $D^{-1}(0, d + f(d)))$ and on which the volume is bounded from below.

Suppose not, and for simplicity let us suppose that there are no local minima. (As in the case of closed geodesics, the existence of a single local minimum by the logical method is no different from the existence of infinitely many.)

According to a theorem of Gromov, $\text{vol}(V) \geq c(n)\|V\|$ for all metrics on any closed n-manifold V with Ricci $\geq -(n-1)$ (and hence for all Riemannian manifolds with $K \geq -1$). Let $v = 500c(n)$.

Let Σ_k be the sequence of homology spheres constructed in chapter 2, section 2.6, theorem 2. We shall consider $M\#\Sigma_k$, and under the assumption that there are finitely many $f(D)$-deep local minima for D on $\mathcal{R}(M)$ shall construct an algorithm to decide which of the Σ_k are spheres and get a contradiction. By Gromov's inequality, if Σ_k is not the sphere, then for any metric on $M\#\Sigma_k$ one knows that $\text{vol}(M\#\Sigma_k) \geq 2v$.

Now let us consider the Gromov-Hausdorff space of Riemannian manifolds with $|K| \leq 1$, vol $= v$, and Diameter $< D + f(D) + 1$. Consider ε chosen as in Cheeger's lemma from the last section, and the effective $\varepsilon/2^n$ cover of this Gromov-Hausdorff space.

Our algorithm now consists of looking for one of two strange possibilities. Start at $M\#\Sigma_k$ (or the element of the net closest to one's initial metric) and begin exploring the nearest neighbors and try to construct a path either to a metric manifold where vol $< 1.5v$ or to M (with its initial metric). If either of these occur, announce that Σ_k is the sphere; otherwise announce that Σ_k is not the sphere.

That this algorithm must work is clear from our hypothesis of contradiction. If $M\#\Sigma_k$ is diffeomorphic to M, then one can connect the given metric to the local minimum M without the diameter ever increasing beyond $D + f(D)$. During such a path one of two things will happen. Either the volume will go below v or it will not. In the first case we will see a path going to a point where the volume is below $1.5v$, and in the second we will find a path all the way to M.

Conversely, if Σ_k is nontrivial, Gromov's inequality together with our inequality on the simplicial norm gives the impossibility of lowering the volume; and, of course, if it is nonsimply connected, then $M\#\Sigma_k$ must have a different fundamental group from M (by Grushko's theorem; see the proof of Novikov's theorem in chapter 2).

This contradiction gives us our components with a lower bound on volume. However, we know that the Gromov-Hausdorff space of manifolds with curvature bounds a lower bound on volume, and an upper bound on diameter is precompact, so we can find a minimum in the closure of each such component. Now we apply the following basic theorem, which can be proven using part of the same analytic package as the smoothing theorem of the previous section (see the notes).

Theorem *The Gromov-Hausdorff limit of a sequence of Riemannian manifolds with curvature bounds, an upper bound on diameter, and a lower bound on volume, is automatically a manifold, and the metric on it is Riemannian of class $C^{1,\alpha}$ for every $\alpha < 1$. (This means that the limit metric is differentiable, with an α-Holder derivative.)*

Now, on to the more quantitative estimates for how many components there are. Our first attempt that gets us close, but not quite there, is essentially geometric:

We would like to produce c^{c^d} doppelganger homology spheres that produce different manifolds from each other, unless they are all the sphere. Consider first a metric sphere of diameter $\sim d$, which has volume c^d; such metrics can be obtained as boundaries of thick regular neighborhoods of triadic trees in hyperbolic space. One can arrange to have $\sim c^d$ little balls connected to one another like beads on a string, where the string has a definite thickness. Now produce the central extension of the group $\pi *_{\mathbb{Z}} \pi' *_{\mathbb{Z}} \cdots$ where a π or π' is placed at each ball, and a $\mathbb{Z}$ whenever the two balls are contiguous on the line. (We will have π and π' as different groups, and the $\mathbb{Z}$'s nonconjugate elements of these groups that we use for making the amalgamations.) This gives a way of encoding an arbitrary binary string of length c^d into a set of homology. (We will have to rescale a bit to get these to have the right bound on the geometry.)

It is relatively easy to build the dichotomy theorems making the π's (if nontrivial) indecomposable with respect to free product over $\mathbb{Z}$, and then, using subgroup theorems for amalgamated free products, arrange that if γ is nontrivial in the nontrivial word problem group then *all of the strings of groups* are nonisomorphic. As a result, if one could get a path from one of these components to another, one would end up learning that γ was trivial.

Unfortunately, this only gives us infinitely many d for which there are this many distinct local minima. To do better, one could build the seed unsolvable word problem group to have nontrivial words of all lengths for which the Dehn function grows very quickly, so the seed π's can be efficiently built. Then one needs to see also that the proof works not only by contradiction, but rather positively, that is, that each doppelganger associated with a curve that is sufficiently hard to shrink to, say, half of its length, actually produces a local minimum. (This is analogous to what we did for the closed geodesic problem in chapter 3, section 3.2.[76])

However, even this runs into a little trouble. We can only use given elements of our seed word problem for so long: at some point we have to increase the size of our seed π. We can use words of length k for $d < BB(k)$. In the end, this argument produces $\exp(\exp(d/BB^{-1}(d)))$. The details are somewhat awkward and lead to a slightly weaker result than we want.

Happily, there is an alternative approach that puts less stress on the geometry and combinatorial group theory and more on the complexity theory, and actually works! One uses the "beads on a string" approach to give an encoding of strings into homology spheres, such that a string lying in Barzdin's set A (see chapter 3, section 3.3) with linearly growing Kolmogorov complexity exactly corresponds to being the sphere. (Note that since we have exponential length strings here, "linear" is precisely what we are looking for!)

The encoding, though, is somewhat different. Rather than a homology sphere "draped over the tree," one can use an acyclic manifold with boundary, and then ensure that it always has the same fundamental group π. (This can be done by putting together basic manifolds with fundamental group according to the tree diagram.) Cross it with a circle C. We take a seed word w that can be one of c^{c^d} possibilities. (At each node of the tree we can put in a different generator of the fundamental group. One can string these choices together because there is a closed path in any triadic tree that visits every vertex and goes through no edge more than twice.) Now attach a 2-handle to the curve (C, w) in the boundary of this product to produce doppelganger homology spheres.[77]

At this point the argument is the same as the one in the closed geodesic problem (see chapter 3, section 3.4). If the algorithm described above ever cut the number of components down too much (it is clearly time bounded), we could use that information to compress *that* initial part of A, defying the complexity bound on segments of A.

4.4 DEPTH VERSUS DENSITY

In this section we shall establish the "fractality" of $\mathcal{R}(M)$, that is, theorem 2 of section 4.1. The proof is a more quantitative reexamination of the proof of theorem 1, globalized over all of $\mathcal{R}(M)$.

---

[76] It requires a step relating the Dehn function of the seed group to the path connecting $M\#\Sigma$ to M within Gromov-Hausdorff space. This can be done by following the isomorphisms of fundamental groups produced by the path, and using effectively proofs of Van Kampen's theorem (how a nullhomotopy of a curve in an amalgamated free product gives a nullhomotopy on one side, assuming injectivity). The explicit estimates are quite unpleasant and related to ones discussed in the next section.

[77] One should modify this construction slightly at one vertex to build in the nontrivial simplicial norm, as we have already done earlier.

The proof is somewhat easier if we assume that the Euler characteristic of M is nonzero. In the algorithm given in the previous section, there were two possible outcomes that correspond to the homology sphere being the sphere: either the volume got very small, or we had a path to M. The first case is somewhat more difficult to analyze explicitly. However, if the Euler characteristic is nonzero, because of the Gauss-Bonnet theorem, one knows that $|K| \leq 1$ implies that $\chi(M) \leq c(n)\text{vol}(M)$, so one can design the algorithm knowing that the first possibility does not occur.[78]

Let us now prove theorem 2. Consider a machine that builds homology spheres according to the Turing machine f (as we described in chapter 3). The first point is that using the disks of given area that kill words, one can build a diffeomorphism of the relevant Σ's to the sphere, so we get an estimate of the length of the path connecting $M\#\Sigma$ to M. Here M is to be viewed as a specific Riemannian manifold, that is, any point on the space we have been calling $\mathcal{R}(M)$. In addition, if one has any estimate of the metric entropy of the part of Gromov-Hausdorff space where our construction takes place, that is, an estimate of the relevant $N(\varepsilon)$, one gets an estimate for the distance from $M\#\Sigma$ to M. These bounds are just iterated exponentials of f itself, and hence f is an adequate bound. This gives the upper bound for the distance one travels to a local minimum.

To get a lower bound on the depth, we want to relate the depth to the Dehn function. To do this, we need to control a few things. The first is (again) the entropy of the moduli space, so as to relate depth to the length of the path that lowers the value of D. This is exactly as in the closed geodesic problem. Then we need to understand quantitatively the homotopy equivalences between ε-close elements of our Gromov-Hausdorff space, and what their norms are on the simplicial chain complexes associated with the covers of the spaces. This is quite easy to do from the explicit arguments of the previous section. Finally, one has to note that after doing all of these maneuvers, the original curve in $M\#\Sigma$ is moved to a curve of some specified length that lies in a disk in M. Thus, the area of the final bounding disk is only a fixed multiple of this final length.

This completes our sketch proof if we have an a priori lower bound on volume. In general, we just have to complicate matters slightly by taking connected sums with some other doppelganger homology spheres which are either the sphere or have nonzero simplicial norm and whose f-time-bounded Kolmogorov complexity is very large, but which can be solved with an $\exp(f)$ bound. (Standard theory enables one to produce sets with this property, and then our encoding results place this set within the context of homology spheres.) Then, one has guaranteed enough candidate doppelgangers with a lower bound on the volume to repeat the above argument.

---

[78]The same reasoning shows that it is enough to assume that the simplicial norm or any characteristic number does not vanish.

Remark. Part of the reason our estimates have been so bad, that is, we have such large towers of exponentials, is no doubt a consequence of our effort to phrase our results in terms of "scales." Note, for instance, that in the closed geodesic problem, one gets out a superexponential Dehn function but only a linear depth local minimum. To get exponentially deep local minima, we would need a super-double-exponential Dehn function, and so on.

In our problem, we can use any Turing machine we want, with any stopping time, thanks to the Birget-Rips-Sapir machinery, and our encodings. Applied to linear depth local minima, we need a tower of a few exponentials to get the existence of a local minimum. In terms of our analysis above, this means the local minima are at least that dense.

Remark. The arguments given here seem to apply to certain 4-manifolds, even ones for which we do not know the unsolvability of the recognition problem. For example, for simply connected 4-manifolds, one can modify by using balanced presentations of the trivial group (i.e., ones with an equal number of generators and relations), which are witnesses to the triviality of words in groups where they are only products of many relators (larger than a large tower of exponentials).

4.5 BDiff

This section is a summary of some known results on BDiff(M), the classifying space of differentiable fiber bundles with fiber M. This space is intimately related to $\mathcal{R}(M)$, and we shall use these results to describe (in principle) the geometry one sees in each basin, at the scale of that basin. For instance, if M has no effective ($=$ nontrivial) smooth group actions, then $\mathcal{R}(M)$ and BDiff(M) have the same homotopy type—the complications caused by group actions will be the subject of the next section.

The analysis of differentiable fiber bundles goes in two steps: first one classifies differentiable "block bundles" and then compares fiber bundles to block bundles. The first is accomplished by means of surgery theory, and the second is the topic of pseudoisotopy theory. (The reader might wish to review chapter 2, section 2.4, appendix 1 for the surgery theory of individual manifolds.)

A block bundle over a polyhedron P with fiber M is a total space E with a map[79] $\pi\colon E \to P$ such that for each simplex Δ of P one has an isomorphism

$$\pi^{-1}(\Delta) \cong \Delta x M.$$

(This isomorphism should be taken as a diffeomorphism of stratified smooth manifolds, the precise definition of which we leave to the reader.) The difference between a block bundle and a fiber bundle is that the isomorphism above does not have to respect the projection map to Δ. This notion is extremely natural in the PL category, but in the smooth category, their introduction is mainly

---

[79] Strictly speaking, the map π is not actually part of the data.

utilitarian as a construct that can be analyzed and can be compared to the objects we are interested in, namely, the fiber bundles.

Block bundles with fiber M have a classifying space, which we shall denote $BD(M)$.[80]

There is a map: $BD(M) \to BAut(M)$, where $Aut(M)$ is the space of self-simple homotopy equivalences of M (this space is just a union of components of the space of self-homotopy equivalences[81] of M). $BAut(M)$ classifies Serre fibrations with homotopy fiber M (with a simplicity condition on monodromy). In any case, the study of Aut is classical homotopy theory.

Let $\mathcal{S}(M)$ denote the homotopy fiber of $BD(M) \to BAut(M)$. According to blocked surgery theory, if $\dim(M) > 4$, there is another fibration which computes the homotopy type of $\mathcal{S}(M)$:

$$\mathcal{S}(M^n) \to \text{Maps}[M; F/O] \to \mathbf{L}_n(\pi_1 M), \qquad (1)$$

where F/O is the classifying space discussed in chapter 2, section 2.4, appendix 1. It is the fiber of the classical J-map studied by the homotopy theorists $BSO(n+1) \to BAut(S^n)$. (BF is the limit of the $BAut(S^n)$ under suspension.) The space $\mathbf{L}_n(\pi_1 M)$ has many fine properties, which are of critical importance in using the fibration (1) in practical calculations, but we will mention only one[82]:

$$\pi_k \mathbf{L}_n(\pi_1 M) \cong L_{n+k}(\pi_1 M) \qquad k \geq 0, \qquad (2)$$

where the groups on the right-hand side are the L-groups of surgery theory.

Using (1) and calculations of L-groups and of self-homotopy equivalence spaces of M, one can often compute the homtopy type of $BD(M)$.

Just as a simple example, if $M = D^n$ is a disk, then the $\mathbf{L}$ term is contractible. So one learns that $\mathcal{S}(D^n) \cong F/O$ (the mapping space is homotopy equivalent to F/O: we are not dealing with base point preserving maps!). So, if one is interested only in rational information, $\mathcal{S}(D^n) \cong BO$. The homotopy groups of $BAut(D^k)$ are the homotopy groups of spheres—the version of Aut for manifolds with boundary is self-maps of the manifold pair, so in this case we would be studying the homotopy groups of the space of self-maps of the sphere—which, as is well known, are finite. One then sees that there is an extra rational piece in $BD(D^n)$, which corresponds to the Euler class of the disk bundle. In short, there is a rational isomorphism $BD(D^n) \cong BO$ for n odd, and $BD(D^n) \cong BO \times K(Q, n)$ for n even.

As another example, if M is closed and aspherical and the Borel conjecture holds, then $\mathcal{S}(M^n) \cong \text{Maps}[M; Top/O]$, where the classifying space Top/O has finite homotopy groups; they vanish until dimension seven, except for $n = 3$ where the group is $\mathbf{Z}_2$. The homotopy groups in dimension

---

[80]The standard notation in the literature is $\widetilde{\text{Diff}}$.

[81]See M. Cohen. *A course in simple homotopy theory.* Springer-Verlag, New York, 1973.

[82]Actually, for the simple homotopy version that we are discussing, it is important to use L-groups with an "s-decoration." For the reader uninterested in decorations, these affect only the groups at the prime 2.

seven and higher correspond to the differentiable structures on the n-sphere. Under these conditions, Aut(M) will be a union of components, corresponding to Out($\pi_1(M)$), each of which is a torus, that can be canonically identified with the classifying space of the center of $\pi_1(M)$. Assuming, for simplicity, that the center of $\pi_1(M)$ is trivial, BD(M) will have the homotopy type of the fibration over BOut($\pi_1(M)$) with fiber Maps[M; Top/O], whose monodromy is associated with the action of Out($\pi_1(M)$) on Maps[M; Top/O] = Maps[B($\pi_1(M)$): Top/O].

Our interest, however, is in BDiff(M).

The connection between BDiff(M) and BD(M) can be analyzed by understanding two stabilization methods.

This analysis begins with the result that our two protagonists closely resemble each other after one crosses with many circles.

$$\begin{array}{ccccccc}
\text{BDiff}(M) & \longrightarrow & \text{BDiff}(M \times S^1) & \longrightarrow & \text{BDiff}(M \times T^2) & \longrightarrow & \cdots \\
\downarrow & & \downarrow & & \downarrow & & \\
\text{BD}(M) & \longrightarrow & \text{BD}(M \times S^1) & \longrightarrow & \text{BD}(M \times T^2) & \longrightarrow & \cdots .
\end{array}$$

The limits of the two lines are homotopy equivalent. (Actually, though, the bottom line is quite easy to analyze, and the maps are injective, away from the prime 2, on homotopy groups. That is not at all the case on the top line, where the map never stabilizes, even injectively.)

So we want to understand what gets lost (or, perhaps better, what changes) when we stabilize by crossing with a circle.

This problem is analogous to what is solved by the s-cobordism theorem. After all, two manifolds are h-cobordant iff they are diffeomorphic after crossing with S^1. The reason is that if $M \times S^1 \cong N \times S^1$, then by unwrapping the circle (i.e., by taking an infinite cyclic cover), one sees M and N both embedded in $M \times \mathbf{R}$. Without loss of generality, these embeddings can be taken disjoint; the region between these is an h-cobordism.

Conversely, any h-cobordism becomes a product after crossing with a circle. (The obstruction to triviality, the Whitehead torsion, multipies by the Euler characteristic when one takes a product.) Thus Wh(π) contains the answer to the question of how manifolds become diffeomorphic after crossing with a circle. So the h-cobordism theorem is about different bundles over a point that become stably equivalent (i.e., after crossing with circles).

It turns out that the problem for higher-dimensional bases involves higher algebraic K-theory. (Note that Wh(π) is a quotient of $K_1(\mathbf{Z}\pi)$.)

As even the defintions in this theory are quite complicated, we will work our way up to the main statements a little bit at a time. For now, let us discuss the first case understood, the one that is most completely and explicitly analyzed in the literature. This case is relevant to the computation of $\pi_0(\text{Diff}(M))$ that is, to analyzing bundles over a 1-complex.

Bundles over a circle are essentially the same as isotopy classes of diffeo-morphisms. Surgery theory gives us directly the information about whether a diffeomorphism is *pseudoisotopic* to the identity, that is, whether $f: M \to M$ extends to a diffeomorphism $M \times I \to M \times I$ which is the identity map on the other boundary component. (After all, we can think of the homotopy as giving us an element of the relative structure set $S(M \times I \text{ rel boundary})$.) Consequently, we are interested in the difference between pseudoisotopy and isotopy. (We will also have to analyze how unique the pseudoisotopy from f to the identity is, to get a complete answer.)

Incidentally, all we have done so far is repeat the discussion above in more concrete words. A bundle over a circle is trivial as a block bundle iff the glueing diffeomorphism is pseudoisotopic to the identity. This analysis extends to block bundles over polyhedra of arbitrary dimension. They are essentially rigged to be analyzable—inductively over the skeleta of the base—by classical surgery.

Thus, one introduces the concordance space $C(M) = \text{Diff}(M \times I)/\text{Diff}(M)$. It turns out that if M is simply connected then $\pi_0(C(M)) = 0$, that is, pseu-doisotopy implies isotopy, and thus surgery tells all about the isotopy classes of diffeomorphisms of a simply connected manifold.

However, if M is not simply connected, then $\pi_0(C(M))$ is never 0, and the exact answer depends on $\pi_1(M)$, $\pi_2(M)$, and the k-invariant that describes the linkage of these homotopy groups with one another. Assuming the k-invariant is trivial, the formula is that (for dim > 5)

$$\pi_0(C(M)) \cong \text{Wh}_2(\pi_1(M)) \times \text{Wh}_1(\pi_1(M); \mathbf{Z}_2 \times \pi_2(M)).$$

Wh_2 is the fiber of the assembly map in algebraic K-theory, and we refer the reader to the paper of Hatcher cited in the notes for the definition of the second term. If $\pi_2(M) = 0$, then the second term is a sum of $\mathbf{Z}_2$'s, one for each nontrivial conjugacy class of element of the fundamental group.

In general there is a contribution from $K_3(\mathbf{Z}\pi_1)$ if the k-invariant is nontrivial.

For example, this formula, and the reasoning above can be combined to give the formula

$$\pi_0(\text{Diff}(T^n)) \cong ([T^n : \Omega(Top/O)] \oplus \oplus \mathbf{Z}_2) \rtimes \text{GL}_n(\mathbf{Z}),$$

where the sum is infinite and indexed by the nontrivial elements of $\mathbf{Z}_n$. The map obtained by crossing with a circle $\pi_0(\text{Diff}(T^n)) \to \pi_0(\text{Diff}(T^{n+1}))$ is actually trivial on all of the $\mathbf{Z}_2$-summands.

Conjecturally, a rather similar formula should be true for any aspherical manifold (of dimension at least six).

In order to study the higher homotopy of $\text{Diff}(M)$, one makes use of the following idea: There is another stabilization process $C(M) \to C(M \times I) \to C(M \times D^2) \to \cdots$ and the maps have a connectivity proportional to the dimensions of the spaces involved. Thus, letting $\mathcal{C}(M)$ denote the limit, the map

$C(M) \to \mathcal{C}(M)$ is highly connected.[83] This limit space is an infinite loopspace and it is what one directly studies. (In fact, and this is a real advantage, by passing to the limit, one can define $\mathcal{C}(X)$ even when X is not a manifold; if X is a finite complex, it measures differentiable concordances on manifold thickenings of X in Euclidean space.)

The form of the calculation goes like this. Waldhausen defines and makes sense out of the expression

$$A(X) = K_0(\mathbf{Z}\pi_1(X)) \times BGL(\mathbf{Z}[\Omega X], \mathbf{Z}),$$

where GL denotes the infinite general linear group. (One should think of $\mathbf{Z}[\Omega X]$ as a generalized group ring.) This is thought to be analogous to the definition of algebraic K-theory given by Quillen, and indeed, for $X = B\pi$, there is a map $A(B\pi) \to K(\mathbf{Z}\pi)$ which is a rational homotopy equivalence.

The main result is that $A(X)$ has $B^2C(X)$ as a summand, and the other summand is $Q(X_+)$, the stable homotopy theory of X together with a disjoint point base point. (Rationally, this second term is homology.)

Combining this calculation with Borel's calculation of $K_*(\mathbf{Z}) \otimes \mathbf{Q}$, one sees that if the isomorphism conjecture discussed in the appendix were correct, one would get for π torsion-free

$$\pi_k C(B\pi) \cong \oplus H_{k-4i-1}(B\pi; \mathbf{Q}) \qquad \text{(conjecturally)}.$$

One actually knows that the right-hand side is a summand of the left-hand side, when $B\pi$ is a finite complex. (There is a much more general conjectural rational calculation that applies when π has torsion; see the appendix.)

The $A(X)$ story does not end here. There has been a great deal of work devoted to extending the above ideas to X that is not aspherical. The idea is to study the deviation of the fiber of the map $A(X) \to A(B\pi)$ from being a homology theory. (It isn't one, unlike the case of surgery theory.) The deviation suggests a "quadratic term" to approximate $A(X)$. (So $A(B\pi)$ is the constant term; the deviating homology theory is the linear term, etc.) Then, the fiber of $A(X)$ to its "quadratic approximation" is also not contractible. So one approximates it by a "cubic functor." In the end, one gets a "Taylor series" that can be used for some calculations.

We will not make use of these extensions; so beautiful and important as they are, we must set them aside. In any case, we shall adopt the rather lofty point of view that $A(X)$ is at least theoretically computable at the level of conjecture.

To summarize, we analyzed block bundles via surgery and homotopy theories, and thus got a description of $BD(M)$. The difference between $BD(M)$ and $BDiff(M)$ seems to be related to the concordance spaces $C(M)$, somehow.

---

[83] Since we will not be discussing the topological version of concordance space theory, we are free to use $\mathcal{C}$ for the smooth concordance space. In the literature, this notation is often used for the topological version, while smooth concordance spaces get blessed with a superscript of "Diff."

We have not said how this is done yet, but we have cited Waldhausen as an authority asserting that $C(M)$ is itself related to algebraic K-theory.

Let us now return to the issue of combining the pieces to get a description of BDiff. As this combination is considerably more difficult to get right at the prime 2, I will leave references regarding how to do this to the notes section, and shall consider only what happens at odd primes (or rationally).

The missing ingredient is the role of Poincaré duality. A-theory in the form we've discussed can be used to describe when a fibration can be modified to be a fiber bundle with compact manifolds with boundary as fibers—after all, it is insensitive to stabilization by crossing with disks. However, for problems involving closed manifolds, there is a nontrivial restriction that is a consequence of Poincaré duality. More precisely, there is an involution on A-theory that reflects the Poincaré duality satisfied by a closed manifold, and algebraically is related to the map $g \to g^{-1}$ on a group, which gives an anti-involution on the group ring. (For a manifold, the algebraic duality gets modified by a sign depending on the dimension.) We shall denote the involution by I.

Just as vector spaces (in characteristic other than 2) can be decomposed into invariant and anti-invariant parts in the presence of an involution, the same can be done for reasonable spaces, provided one localizes away from the prime 2. We will use the notations $C(M)^{\pm I}$ to denote these pieces.

What happens is that the fiber of the map $\mathrm{BDiff}(M) \to \mathrm{BD}(M)$ is essentially $C(M)^{\pm I}$ (where the $\pm$ is determined by dimension). This means concretely that there is a long exact sequence whose homotopy groups are the (anti-)invariant parts of the homotopy groups of the concordance spaces.[84,85]

That this should enter is not mysterious. Let us return to thinking about the difference between block bundles and bundles; over 1-simplices, we will have, in the block bundle case, h-cobordisms between M and itself. The torsion of such an h-cobordism is not arbitrary. It has a self-duality condition[86] (which differs up to sign depending on dimension) which reflects the fact that the

---

[84]It is easy to see why that could be the case by thinking about h-cobordisms. A necessary condition for the boundary components to be the diffeomorphic (by a diffeomorphism in the homotopy type of the inclusions) is that the torsion of the h-cobordism be (anti-)self-dual. Every torsion of the form $x + Dx$ $(x - Dx)$ can be produced by doubling along a boundary component an h-cobordism with torsion x, and therefore does not have both boundary components diffeomorphic. However, if 2 is inverted then any self-dual x is of the form $y + Dy$ by setting $y = x/2$, and the necessary and sufficient conditions coincide.

[85]Technically, what one really needs is the Tate homology of $\mathbf{Z}_2$ acting on $C(M)$. Alas, there are serious technicalities involved in developing the homological algebra of spectra.

[86]Intuitively, Whitehead torsion has formal similarities to the torsion in the homology of a chain complex of a rationally acyclic complex. If the complex satisfies duality in dimension n, the torsion has a duality linking dimensions that are $n - 1$ complementary. Poincaré duality then gives a vanishing when n is even, and no condition for n odd, which is what the duality condition means in this case.

composite of the inclusion of one boundary component with the retraction to the other is a simple homotopy equivalence.

Thus, in a stable range proportional to the dimension of the manifold (here the difference between unstable and stable concordance spaces must enter), BDiff is determined as a mixture of the self-homotopy equivalences, L-theory, and algebraic K-theory (of spaces). While the details are evidently quite complicated, it might at this point be imaginable or believable to the reader that in special cases, such as aspherical manifolds for which all sorts of Borel type conjectures are known, one could make some calculations, at least in the stable range.

What is of most direct interest to us is the realization that fundamental group homology enters into this description via both the K-theory and the L-theory. By making use of the techniques of chapter 1, we can arrange to have these homology groups as rich as we would like for doppelgangers of arbitrary manifolds.

Appendix 1: The Isomorphism Conjecture and Secondary Invariants

Earlier we discussed the Novikov and Borel conjectures; these are very fine conjectures that give us a great deal of help in guessing what $S(M^n)$ looks like when $\pi_1(M)$ is torsion-free. Indeed, there should be a fibration

$$\text{Maps}\,[M; Top/O] \to S(M^n) \to \text{Maps}\,[B\pi/M; B(F/Top)].$$

We think of M as a subset of $B\pi(M)$, as usual, by taking a mapping cylinder. (The maps in this sequence are somewhat subtle: $\text{Maps}\,[M; Top/O] \to S(M^n)$ is produced by "smoothing theory." The map $S(M^n) \to \text{Maps}\,[B\pi/M; B(F/Top)]$ requires a proof of the Novikov conjecture to define. After all, there is a restriction map

$$\text{Maps}\,[B\pi/M; B(F/Top)] \to \text{Maps}\,[M; F/Top].$$

The statement that we can lift the map $S(M^n) \to \text{Maps}\,[M; F/Top]$ of the surgery sequence to $S(M^n) \to \text{Maps}\,[B\pi/M; B(F/Top)]$ implies that there is a cohomological restriction, depending on $\pi_1(M)$, on the part of $\text{Maps}\,[M; F/Top]$ coming from homotopy equivalences. This is essentially the content of the Novikov conjecture.)

It is important to have a similar conjecture to guide us when π contains torsion.[87] We note that we can obtain results true for all manifolds from partial verification of conjectures: what one needs to do is build in the hypotheses of these partial verifications in the constructed doppelgangers.

Note that any such conjecture should treat finite groups specially, simply because we know a great deal about their L-theory and enough about their

---

[87] As everyone knows, conjectures need not be true to be valuable.

K-theory to see that these are very complicated and directly connected to arithmetic. We would be setting ourselves up for a lot of trouble if we insisted on describing arithmetic geometrically.[88] Our first idea is to try to compute our functors in terms of the L-theory of the finite subgroups.

One way to motivate this is to consider the parenthetical remark made in our first discussion of the Borel conjecture. If we think of the Borel conjecture as a version of the topological rigidity of closed spherical manifolds, we can try to consider what the topological rigidity of orbifolds, say, quotients of symmetric spaces by cocompact proper discontinuous isometric discrete group actions, should mean.

There is an h-cobordism statement, which we will not be explicit about, and there is a surgical component. Let G act nicely, for example, cocompactly and properly discontinuously, on the symmetric space X^n.

We can think of X/Γ as a stratified space. We can now invoke the stratified analogue of surgery theory, which tells us when stratified homotopy equivalences are stratified homotopic to homeomorphisms. In this theory, the L-homology term in the ordinary theory becomes replaced by a more (co)sheaf theoretic object that takes into account the variation of the local structure of the stratified space. Then, just as in the manifold case, we can assert that there is a map

$$H_n(X/\Gamma; \mathbf{L}(\Gamma_x)) \to \mathbf{L}_n(\Gamma),$$

which (assuming, inductively, the rigidity of all of the lower-dimensional strata) could be used to algebraicize the assertion that X/Γ is topolgically rigid. Rigidity would be the statement that the map is an isomorphism. (The h-cobordism statement would assert a similar thing with algebraic K-theory replacing L-theory.)

Again, note that when all the isotropy is trivial (i.e., the action is free, or, equivalently, Γ is torsion-free), then all of the $\mathbf{L}(\Gamma_x)$ become the same coefficients $\mathbf{L}$, and we have the usual statement of the Borel conjecture.

Alas, this statement is false. In algebraic K-theory this failure goes back to the very beginnings of the subject. The "fundamental theorem" of algebraic K-theory asserts that

$$K_n(\mathbb{Z}[\mathbb{Z} \times \pi]) \cong K_n(\mathbb{Z}[\pi]) \times K_{n-1}(\mathbb{Z}[\pi]) \times \mathrm{Nil}_n(\mathbb{Z}[\pi]).$$

Nil is an extra unexpected (and unwanted!) piece from our point of view. (Strictly speaking, there are two Nil terms, but we shall ignore these. For us, Nil just means "the unknown.") In L-theory, the example is more recent, but it is due to a functor Unil analogous to Nil. The most concrete case arises for $\Gamma = \mathbb{Z}_2 * \mathbb{Z}_2$, the infinite dihedral group which acts on the real line. In that case, the extra piece is an infinitely generated 2-group.

---

[88] Actually, the complications are all at the prime 2: away from 2, the L-theory of finite groups can be directly explained by representation theory; similarly, the isomorphism conjecture can be simplified away from 2.

There are two ways around this problem. The simplest is to invert 2 (for L-theory) which kills the Unil problem. This seems to work, as far as we know. Of course, we have avoided the problem by throwing away deep significant aspects. And, in practice, a good many of the most interesting questions boil down to 2 local phenomena. (Yes or no is a $\mathbb{Z}_2$ question.) On the other hand, it suffices for our applications to secondary invariants.

The second approach is deeper. One views the original Borel conjecture as the attempt to compute $L(\Gamma)$ in terms of the trivial group using some sort of functoriality. ($B\Gamma$ can be viewed as the classifying space of the category with one object, all arrows isomorphisms, and the groupoid of arrows coinciding with Γ.) Alternatively, one can think of $B\Gamma$ as the terminal object in the setting of spaces which have maps of their fundamental groups to Γ. Now, using all of the maps of all the balls in (a Leray cover of) $B\Gamma$, and making use of the functorialities present with respect to inclusions, one gets a version of the left-hand side.

The attempt we described above is similar, except it explicitly takes into account the finite groups. In geometric terms, we are using spaces with proper Γ-actions.

The existence of the Nil term in the fundamental theorem and the example of the dihedral group in L-theory assert that funny things occur for groups that are *virtually cyclic*.[89] As a result, one wants a variant of the left hand side that explicitly takes into account virtually cyclic groups.

The way one does this is as follows[90]: One introduces a universal space X on which Γ acts cellularly, such that for all virtually cyclic subgroups Δ of Γ the fixed set X^Δ is contractible. (Note that, in particular, X is.)

Using this one then conjectures that

$$H_n(X/\Gamma; \mathbf{L}(\Gamma_x)) \to \mathbf{L}_n(\Gamma)$$

and

$$H_n(X/\Gamma; \mathbf{K}(\Gamma_x)) \to \mathbf{K}_n(\Gamma)$$

or the versions with C are all isomorphisms. (Note that now one can also work purely algebraically and use coefficients in a general ring R, etc.) The only thing that changed is the X! Remarkably, this conjecture has been verified in a great many cases.

Just to give a simple special case, let us assume that Γ is nonpositively curved and torsion-free. Then the only virtually cyclic subgroups are isomorphic to $\mathbb{Z}$. Every cyclic subgroup is contained in a unique maximal cyclic group (these correspond to the closed geodesics). One can analyze the left-hand side by comparing to the Borel construction $X \times_\Gamma E\Gamma$, which is homotopy equivalent to $B\Gamma$. One discovers that the difference between $H_n(X/\Gamma; \mathbf{K}(R\Gamma_x))$ and

---

[89]A group is virtually cyclic if it contains a cyclic subgroup of finite index.

[90]For those familiar with category theory, another convenient way to describe this is as the left Kan extension of the functor defined on the subcategory of spaces whose fundamental groups are virtually cyclic.

$H_n(X/\Gamma; \mathbf{K}(\Gamma_x))$ is concentrated along the singularities, which are in a 1-1 correspondence with the closed geodesics (these maximal $\mathbb{Z}$'s). For each of these, one gets a $\mathrm{Nil}_n(R)$. Other considerations let one know that these actually split off the K-theory. For $\mathcal{C}$, the same result holds. (Note that $\mathcal{C}(S^1)$ is quite different from Maps $[S^1 : \mathcal{C}(*)]$.)

This completes our discussion of what we conjecturally believe K-theory and L-theory of group rings to look like. We will spend the rest of this appendix describing some of the implications of this picture to the understanding and construction of some invariants of manifolds.

Let us first look at what surgery theory says about the classification of manifolds with finite fundamental group.

First, the simply connected case. The answer is quite simple: $\mathcal{S}(M^n) \otimes \mathbb{Q} \cong H_{n+1}(*, M; \mathbf{L}) \otimes \mathbb{Q}$. (This concretely means that a manifold is determined within its homotopy type by its rational Pontrjagin classes, up to finite indeterminacy, and that the only restriction on these classes is that forced by the Hirzebruch signature theorem.)

In the nonsimply connected case (remember π is finite for now), one has to contend with $L_{n+1}(\pi)$. Rationally, this is quite easy to understand: it vanishes when $n + 1$ is odd, and is isomorphic to some representations when $n + 1$ is even. (Actually, L_{2k} is made up out of $(-1)^k$ symmetric bilinear forms, and by using representations of π into finite-dimensional vector spaces one can get invariants of these quadratic forms.)

This part can be analyzed a priori in a couple of different ways. There is an analytic approach, using spectral theory, that I shall not discuss. However, for topologists, the simplest method is to make (any manifold homotopy equivalent to) M bound a manifold with fundamental group π,[91] and then consider the action of π on the cohomology of the universal cover.[92] It gives exactly the right kind of quadratic form. When one mods out by the indeterminacy that one sees already on a closed manifold, one gets exactly an invariant that can detect $L_{n+1}(\pi, e)$. This invariant (depending on which definition one uses, and how one parametrizes it) is called either $\rho(M)$ or $\eta(M)$. In any case, the result for finite fundamental group is that the determination is made on the basis of Pontrjagin classes (subject to the Hirzebruch restriction), together with the ρ invariant.[93]

---

[91] Strictly speaking, one must take the disjoint union of several copies of M in order to accomplish this.

[92] If one had to take n copies of M to produce the bounding manifold, then here one should divide by n.

[93] Actually, what I described is the classification of manifolds within a given simple homotopy type. If one wants to go to the homotopy type, there is another invariant of the homotopy equivalence, the Whitehead torsion, which played an important role in our discussion of BDiff. This torsion must be "anti-self-dual" in even dimensions, and "self-dual" in odd dimensions. This part of the calculation is exactly analogous to the way one must combine the concordance space information with block-bundle surgery information to get information about BDiff.

For manifolds with infinite fundamental group, the story is more complicated. The first piece of the modification is straightforward. Using a solution to the Novikov conjecture (if there is one),[94] one can define a map

$$S(M) \rightarrow H_{n+1}(B\pi_1(M), M; \mathbf{L}) \otimes \mathbb{Q}$$

(which turns out to always have as image a lattice within this $\mathbb{Q}$ vector space). This reflects the way in which the higher signatures of M are homotopy invariant.

Already, though, this is somewhat less satisfactory in that this is no longer a difference between intrinsic invariants of the manifolds. If we believe the Borel conjecture and π is torsion-free, this refined higher signature really does tell the whole story. However, if π has torsion, then we are missing a great deal more, and we need an extension of the ρ-invariant to capture it.

As they say, there's good news and bad news. The bad news is that there can be no invariant that does this in general because there are cases where one can compute the effect of composing with a self-homotopy equivalence of M and see that the L-theory piece is directly affected. In particular, the action of the L-group on the structure set cannot be detected—in general—by a difference of intrinsic invariants.

The good news is that there are several different settings where one can get around this. Here is the simplest, namely, the "antisimple case." We shall omit the details, giving only the gist of the construction. It is only the spirit that will be necessary to follow the applications in section 4.7; a more detailed discussion can be found among the notes.

The background for this is a definition of L-groups (due to Mischenko, and refined at 2, by Ranicki) as a cobordism group of chain complexes that satisfy Poincaré duality. An element of $L_n(R)$ is a chain complex with a chain equivalence to its dual, shifted by n. The basic example of such a thing, for R a group ring, is given by the chain complex of a manifold with fundamental group mapping to π. (The chair complex is the cellular chain complex of the regular cover associated with this homomorphism.) The element is trivial if it is "nullcobordant," by a definition that apes the situation one sees for a compact manifold with boundary.

We suppose that the cochain complex of M, $C^*(M)$, is chain equivalent (as a complex of $\mathbb{Q}\pi$ modules) to one with vanishing groups in the middle dimension. This implies that the class of $L_n(R)$ represented by M is trivial; one can just write down an explicit algebraic coboundary. This class is called the *symmetric signature* of M. It turns out that this vanishing implies (assuming the Novikov conjecture) that M bounds a nice stratified space[95] with fundamental

---

[94] A "solution to the Novikov conjecture" is a technical concept; every solution to the conjecture implies the existence of a "a solution to. . ."; we leave aside any discussion of functoriality of such invariants (and solutions).

[95] Called a Witt space.

group π, nice enough for one to be able to talk about associated self-dual inner product pairings and the like. The reason for this is that the Novikov conjecture asserts that L-classes are detected in L-theory, so the above vanishing gives one the vanishing of enough characteristic numbers to get some sort of geometric coboundary for our manifold. (That one can get one with all the right properties is a happy consequence of intersection homology theory.) The "difference" between the algebraic and geometric coboundaries, up to indeterminacy in the cobounding space, defines an invariant.

If one believes in the isomorphism conjecture (any form of it), and one works rationally so that one can use a "Chern character" description of the left-hand side, then one gets an invariant:

$$H\rho(M) \in L_{n+1}(\pi)/ \oplus H_{n+1-4i}(B\pi; \mathbb{Q})$$
$$\approx \oplus H_{n+5+4i}(B\pi; \mathbb{Q}) \oplus [\oplus H_{n+1\pm 2i}(B\mathbb{Z}(g); \mathbb{Q})]^{\pm}.$$

The second sum is over conjugacy classes of g of finite order, $\mathbb{Z}(g)$ denoting the centralizer of g; we will not give the indices. The superscript of $\pm$ is related to a $\pm$ symmetry condition determined by the involution sending g to g^{-1}. (If we were dealing with $\mathbb{C}\pi$, then there would be no condition; the trouble is that we are dealing with real representation theory here, so there is a condition.)

Thus, for instance, if π is torsion-free, then for M with fundamental group $\pi \times \mathbb{Z}_2$, $H\rho(M) \in \oplus H_{n+5+4i}(B\pi; \mathbb{Q}) \oplus H_{n+1\pm 4i}(B\pi; \mathbb{Q})$, the first piece associated with the identity element and the second with the element of order two. If we were in a larger group with an involution in it centralized by π, the first term would be approapriately modified, and the second would remain the same. (For $\pi \times \mathbb{Z}_3$, the calculation would work out as $H\rho(M) \in \oplus H_{n+5+4i}(B\pi; \mathbb{Q}) \oplus H_{n+1\pm 2i}(B\pi; \mathbb{Q})$.)

To get an invariant in $\oplus H_{n+5+ri}(B\pi; \mathbb{Q}) \oplus [\oplus H_{n+1\pm 2}(B\mathbb{Z}(g); \mathbb{Q})]^{\pm}$ one does not need the whole isomorphism conjecture; one needs the rational injectivity of the map that underlies it. This is known in a great many cases.

Surgery theory can be used to show that this invariant can be realized (more precisely, as one varies the elements of $S(M)$, one fills a lattice in that vector space).

There is an analogous idea in concordance theory, but one needs genuine acyclicity to define the invariants.

I will close with two final remarks about $H\rho(M)$. The first is that one does not need M itself to be antisimple. If we have $B\pi$ a Poincaré complex and a (nonzero degree) map $M \rightarrow B\pi$, we can ask that the relative chain complex be anitisimple, and this will be enough. In general, one can define relative invariants, but those have the usual disadvantages, and I will try to avoid them.

The second remark is that one can define similar invariants for families, that is, fiber or even block bundles, with antisimple fibers. The basic point is that one needs a trivialization of symmetric signature of the total space of such a bundle. This is not purely a tautology: the total space is typically not itself antisimple, and there are examples where the symmetric signature is actually nonzero, so strictly speaking this is not even true. However, away from the prime 2 one can prove this—we leave further discussion for the notes.

Appendix 2: JSJ Decompositions

Because automorphisms of manifolds induce outer automorphisms of their fundamental groups,[96] and all invariants that we consider are so dependent on the fundamental group, it behooves us to understand Out. One very useful approach is via canonical decomposition of π.

The first basic fact, which follows from the work of Kurosh and Grushko (hinted at in chapter 1), is that every finitely generated group can be expressed as a free product of freely indecomposable groups. The free summands can be mixed up like crazy, but the remaining ones can be mapped only to conjugates of each other—so if these are different, then Out breaks up as a product of the individual Outs.

Rips and Sela have pioneered the search for theorems describing all amalgamated free product decompositions of a given group where the amalgamating subgroup is small. The prototype for such a theorem is the characterstic submanifold theorem for 3-manifolds, due to Jaco-Shalen and Johansonn; it corresponds to the case where π is the fundamental group of a 3-manifold, and the amalgamating subgroup is abelian (which is necessarily $\mathbb{Z}$ or $\mathbb{Z}^2$ in that case). Notice that, just as in the case of a trivial amalgamating subgroup there are exceptional groups where there are many decompositions, which "hang off" the main group, the same is true in the case of 3-manifolds (the Seifert part of the geometric decomposition of the manifold) and in the more general versions of the theory. In the Rips-Sela theory, the amalgamating subgroups are virtually cyclic (i.e., have a cyclic subgroup of finite index) and the sources of exceptional behavior are quadratically hanging subgroups.

Unfortunately, the JSJ decomposition is not entirely unique. Given any π-action on a tree with virtually cyclic edge stabilizers, there is an equivariant map to the JSJ action. Usually, this gives rise to a uniqueness, up to various conjugations and a "slide move." Forester analyzed this completely: In terms of graphs of groups, the requirement for uniqueness is that at each vertex no two neighboring edge groups are comparable (under inclusion).

I will also mention several other JSJ decompositions: Dunwoody and Sageev, and Fujiwara and Papasoglou have all given versions with other small groups as the stabilizers. (Forester's theorem applies to these, as well.) Scott and Swarup have given a coarser JSJ decomposition, which has the virtues of always being unique and also restricting to the canonical JSJ decomposition for 3-manifolds (which the others do not).

---

[96] Recall that the outer automorphism of a group π is $\mathrm{Aut}(\pi)/\mathrm{Inn}(\pi)$, where Aut is the group of self-isomorphisms of π (with respect to composition) and $\mathrm{Inn}(\pi)$ is the normal subgroup of inner automorphisms of π; $\mathrm{Inn}(\pi)$ is isomorphic to $\pi/Z(\pi)$, π modulo its center.

Finally, Mosher, Sageev, and Whyte have given essentially orthogonal conditions for when amalgamated free product decompositions can be coarsely determined. These can be applied, for instance, when the amalgamating subgroup has low cohomological dimensional and the vertex groups are higher-dimensional Poincaré duality groups (or have suitable vanishing conditions). While the succeeding pages will be written using the more traditional JSJ decompositions, it would not be very hard to modify them to make use of the MSW theory instead.

E. Rips and Z. Sela. *Canonical JSJ decomposiitons, cyclic splittings of finitely presented groups and the canonical JSJ decomposition.* Ann. Math. (2) 146 (1997), no. 1, 53–109.

M. Forester. *Uniqueness of JSJ decompositions of finitely generated groups.* Comm. Math. Helv. (to appear),
http://www.maths.warwick.ac.uk/~forester

M. Dunwoody and M. Sageev. *JSJ-splittings for finitely presented groups over slender groups.* Invent. Math. 135 (1999), no. 1, 25–44.

K. Fujiwara and P. Papasoglou. *JSJ decompositions and complexes of groups.* 1996 preprint.

P. Scott and G. Swarup. *Regular neighborhoods and canonical decompositions of groups.* http://front.math.ucdavis.edu/math.GR/0110210

L. Mosher, M. Sageev, and K. Whyte. *Quasi-actions on trees.* Research announcement, http://front.math.ucdavis.edu/math.GR/0005210

4.6 THE CONTAGION OF SYMMETRY

As explained at the outset of the previous section, $\mathcal{R}(M)$ differs homotopy theoretically from $\text{BDiff}(M)$ because of smooth compact group actions on M. When looked at from this point of view, we quickly discover how little is really understood about the theory of compact group actions, despite its many years of study. It is extremely rare to have very complete understanding of all of the group actions on a manifold.

What we will focus on in this section is that it is extremely rare that to find $\mathcal{R}(M)$ singular, but with only a few singularities. Very typically, if a manifold has some symmetry, it has infinitely many. (These give us complicated filling functions for $\mathcal{R}(M)$ based on singularities.)

In the reverse direction, we will also mention a couple of theorems that eliminate the possibility of symmetry on certain manifolds. (This will give us large singularity-free regions in $\mathcal{R}(M)$.)

Definition. If G acts smoothly on a manifold M, then the singular set $\Sigma(M)$ of M consists of all points such that the isotropy of M is nontrivial.

Theorem *If G acts smoothly on a manifold M such that the singular set of the action $\Sigma(M)$ is a proper subset of M and has at least three-dimensional orbit space, $\dim(\Sigma(M)/G) > 2$, then there is are infinitely many nonconjugate G-actions on M, distinguished by the stratified diffeomorphism type of $\Sigma(M)/G$.*

It is, of course, relevant to us that if $\dim(\Sigma(M)/G) > 4$, then one cannot algorithmically decide whether these actions are conjugate.

The proof is to change a stratum of the singular quotient by taking the connect sum with a homology sphere; this modification can be extended to all the strata "above" it, by using a contractible manifold that the homology sphere bounds. (See chapter 2 for the relevant embedding theory for homology spheres that this modifies.) Extending to all of M/G produces a slightly modified space, which I will call "M/G." "M/G" has a stratified map to M/G, which is an integral homology equivalence on all strata. The pullback of the stratified system of bundles over M/G that produces M with its G-action produces a new G-manifold with quotient "M/G." A little reflection shows that the manifold supporting the action is M.

(The above might have gone too fast. Here is the same proof in a special case. Suppose we start with the circle action on S^n with S^{n-2} as its fixed set. The quotient space is a disk D^{n-1}. The modification under discussion is to consider another contractible manifold with boundary some homology sphere Σ^{n-2}. We can now "spin" this contractible manifold around its boundary to obtain a new manifold, provably diffeomorphic to S^n, and with a circle action whose fixed set is Σ^{n-2}.)

There are many other ways to modify an action. The following applies to another extreme, when there is no singular set at all:

Theorem *If G is a finite group which acts freely on a simply connected closed manifold M of dimension $4k + 3$, then G acts freely on M in infinitely many nonconjugate fashions. (Moreover, the same is true if $\pi_1(M/G)$ is isomorphic to $\pi_1(M \times G)$.)*

This is easily done using surgery theory and the ρ invariant. (For $4k + 1$ this result is true for some groups and not others.)

In the situation of this theorem, one can often determine algorithmically this conjugacy type (e.g., if M is simply connected). However, if $\oplus H_{n+1\pm4i}(\pi_1(M); \mathbb{Q})$ is not a computable group (see chapter 1), then one often cannot determine conjugacy, at least up to conjugacy by diffeomorphisms homotopic to the identity, for $H\rho$ reasons. Since one cannot tell whether $\pi_1(M)$ has this property for appropriate doppelgangers of M, the logical method can be combined nicely with these observations.

There are a number of other ways of "deforming" actions (deformation is a rather inappropriate word, since the set of actions is always discrete!), which can be used to give homotopical information about $\mathcal{R}(M)$, and show the phenomenon of "proliferation of strata," but which do not lead to geometric information beyond the homotopy theory (as far as I know).

1. Changing the equivariant tangent bundle of the action. The G-action on M gives rise to an unstable tangent bundle for the underlying G-space. One can try to modify this. The coarsest feature to be modified would be (for example) the normal representation to the fixed-point set. Doing this really changes the nature of the type of bundle theory one has: the normal bundle to the fixed set will split up into eigenspaces (or more accurately, according to the irreducible representations of G), and the number of these and their dimensions determine different unstable classical groups as the structure group of this bundle. (Another way to say this is that the equivariant version of $O(n)$ has many components, with rather different homotopy types.) When things work out, this can be analyzed by versions of surgery theory.[97]

2. When the difference of dimensions is not high enough, then the fixed set can be re-embedded in a different way. In the PL and Topological categories, it is known that these different embeddings can often also be made into fixed-point sets. I do not know whether or not a similar principle holds in the smooth category (again, except in very special cases).

3. For p-groups, the various strata of the singular set can often be replaced by mod p homology equivalent manifolds. If a suitable equivariant normal bundle can be found for this new singular set, then the technique of "homology propagation" can be used to produce a new action, if suitable obstructions vanish.[98]

4. For non-p-groups, the situation is much wilder and not very well understood. It is only quite recently that the complete analysis of the possible fixed sets of smooth G-actions on *some* disk was accomplished. It is a mixture of a congruence condition on the Euler characteristic of the putative fixed-point set (the modulus of this congruence is called the Oliver number of G) and some bundle theory (e.g., for odd-order cyclic groups, the normal bundle needs a complex structure).

---

[97] However, it is worth noting that they can be rather complicated to work with because of the unstable homotopy theory that arises. This can occur even if a "strong gap hypothesis" applies, because even if the codimension of the fixed set is large, the dimension of the part of the normal bundle associated with a particular irreducible representation can be small. On the other hand, when one is dealing with a cyclic group, and these summands have one complex dimension, then the instability of the homotopy theory is a blessing: one gets a product of $U(1)$'s, whose unstable homotopy theory poses no difficulties.

[98] In the literature this is done under an assumption of "homological triviality." It can also be accomplished under a suitable "gap hypothesis." The obstructions are rather better understood in PL and Top, but they can sometimes be managed in the smooth category. For group actions on nonsimply connected manifolds, the theory of computations is much weaker. It would be a worthwhile project to study this problem, assuming that the isomorphism conjectures discussed in appendix 1 (to section 4.5) hold.

Unfortunately, the above are only partially analyzed: at times it feels more like art than science. We will now turn to some homological criteria that are simple to apply and lead to *nonexistence results* for group actions.

Theorem *Suppose that the center of $\pi_1(M)$ is trivial and that M represents a nontrivial rational homology class in $H_n(B\pi_1(M); \mathbb{Q})$, then every connected group action on M is trivial.*

Without the condition on the center, the conclusion is that the G-action factors through an action of the torus with finite isotropy, and that the latter defines an embedding of $\mathbb{Z}^k$ in the center of π, through the associated orbit embedding of the torus in M.

The hypothesis of the theorem holds, for instance, whenever the simplicial norm of M is nontrivial.

Theorem *Suppose that $\pi_1(M)$ is centerless and, for example, contains no elements of order p, and that M represents a nontrivial element of $H_n(B\pi_1(M); \mathbb{Z}_p)$, then any p-group action on M induces a faithful representation $G \to Out(\pi_1(M))$.*

Consequently, if M represents an indivisible element of $H_n(B\pi_1(M); \mathbb{Z})$ then any effective G-action induces a faithful representation of $G \to Out(\pi_1(M))$.

Note that the second theorem implies the first in the torsion-free case when the homology of π is finitely generated. For the general case, one must modify the proof of the theorem rather than merely apply the result. In a nutshell, the idea is that if g is an element of the kernel of this representation, then it generates a cyclic group C, for which the fundamental group of the Borel quotient $M \times_C EC$ is $C \times \pi$. In particular, since M/C has a map to $B\pi$ (obtained by comparing to the Borel conjecture), by considering the orbit map $M \to M/C$, one gets the equation for homology classes: $[M] = |C|[M/C]$ in $H_n(B\pi_1(M); \mathbb{Z}_p)$, which contradicts the hypothesis.

We note that one does not really need a torsion-free fundamental group; one can mod out by the normal subgroup generated by $\mathbb{Z}/p$; we will, when we need it, easily produce doppelgangers that have torsion-free quotients of the fundamental group that are "sufficiently characteristic" to be used to eliminate group actions.

4.7 FILLING FUNCTIONS FOR $\mathcal{R}(M)$

Now we have enough background to prove the remaining theorems mentioned in section 4.1.

We can prove theorem 3 for low values of k (e.g., $k < n/6$ for many manifolds) by proving a dichotomy theorem of the following sort:

Proposition   *For every manifold M which is relatively antisimple with respect to a $K(\pi, 1)$ manifold, it is the case that, for every $k < n/6$, there is a computable sequence of manifolds M_a for which $H_k(\mathcal{R}(M_a); \mathbb{Q})$ has arbitrarily large rank, and for which there is no algorithm to decide whether $M_a \approx M$.*

The idea is to use the methods of producing large group homology from chapter 2, and realize them as values of $H\rho$ for block bundles over a sphere. Then the material of the previous section realizes these classes in BDiff, and fairly routine arguments apply to homology as well as homotopy. Finally, one must check that these results apply to $\mathcal{R}(M_a)$ and not just to BDiff(M_a)—that will be explained below. Essentially, one produces "asymmetric doppelgangers."

It is not hard to prove a stronger theorem that, in addition to leading to growing images of homology, also provides for kernels of (at least) specific sizes occurring in specified locations.

However, to prove the theorem for all k and all M, it is easiest to use the JSJ theory directly. (This will produce many cycles; however, unlike the previous method they are not guaranteed to be in the image of the Hurewicz homomorphism. As you will see, the representing cycles are tori, not spheres.)

Proposition   *For every manifold M of dimension > 4 and for every k, there is a computable sequence of manifolds M_a for which $H_k(\mathcal{R}(M_a); \mathbb{Q})$ has arbitrarily large rank, if $M_a \not\approx M$, and for which there is no algorithm to decide whether $M_a \approx M$.*

Here the method is to produce a split surjection Diff(M_a) $\to \mathbb{Z}^r$ for arbitrarily large values of r. The map to this abelian group comes from the JSJ decomposition, and the splitting will be via "Dehn twist diffeomorphisms" that are made available by a specific construction of the doppelgangers. Taking "B" gives us large tori in BDiff(M_a)'s.

More specifically, we begin with a slight extension of the classification of fundamental groups of homology spheres (provable by the same method).

Proposition   *If G is a finitely presented group with $H_1(G) = H_2(G) = 0$, and M is any compact manifold of dimension > 4, then there is a manifold M' and map $M' \to M$, which is a $\mathbb{Z}\pi$ homology equivalence, and $\pi(M') = \pi \times G$.*

We can now replace our fundamental group (by a doppelganger) with fundamental group $\pi \times G$, for a group G that is hard to trivialize. This ensures that in our dichotomy theorems the alternative is a freely indecomposable group with elements of infinite order, and can be assumed to have an unsolvable word problem.

The next change in our construction of doppelgangers is to "modify along curves instead of around points." In other words, rather than removing a regular neighborhood of a point (i.e., a ball) and gluing in a homology ball, we shall remove the tubular neighborhood of a curve, and glue in a homology circle, and, even better, a knot complement.

This has several advantages. First of all, we have arranged, assuming the original curve is not nullhomotopic, to have a nontrivial JSJ decomposition. If we assume that the complement has a trivial decomposition, then we even know the JSJ decomposition of this new manifold. We can then repeat this along other curves in either M or the knot complement over and over, and get a complicated treelike construction of doppelgangers. Note that each step of the construction gives us an $S^2 \times S^{n-2}$ embedded in codimension one. There is a "Dehn twist" supported in a neighborhood of each of these. Since these have disjoint support, they commute. If the knot complements we used are sufficiently asymmetric, the JSJ theory shows that thes Dehn twists comprise Out of the fundamental group of the graph sum of the knot complements. Since there is a map from Out(π) to Out of any part of the canonical JSJ decomposition, we see that we have split off this free abelian group, completing the construction.

To complete the proof of theorem 1 and to prove the remaining theorems, it is necessary for us to have a homotopical model for $\mathcal{R}(M)$. We shall use the standard idea of Borel. Consider BDiff$(M) = $ Riem$(M) \times_{\text{Diff}(M)}$ EDiff$(M) \to \mathcal{R}(M)$. The fiber over a metric space (M, g) is Biso(g) where Iso(g) is the (compact Lie) group of isometries of the metric space (M, g).

Given a compact group of diffeomorphisms, G, there is a metric invariant under this group, and "usually" the metrics that are preserved by G are not invariant under any larger group.

The open stratum that contains an element (M, g) whose isometry group is G is abstractly homotopy equivalent to BNorm$(G)/G$. Note that Norm(G) is not the group of equivariant diffeomorphisms of M with its G-action. It also contains elements which preserve the orbits but "permute" the elements of G that are acting. For instance, the reflections on a circle lie in the normalizer of the circle acting on itself by rotation. The succinct way to summarize this is then

$$H_{k+1}(\mathcal{R}(M), \text{BDiff}(M)) \approx \text{hocolim}(H_*(\text{BDiff}(M), \text{BNorm}(G)),$$

where the hocolim is taken over the category of conjugacy classes of compact subgroups of Diff(M). Here are a few special cases worth noting:

If all compact groups acting are finite, then $\mathcal{R}(M) \approx$ BDiff(M) rationally.

If all the positive-dimensional group actions are conjugate to actions of subgroups of a given compact group action, then there is a terminal object in this category, and the right-hand side boils down to simply BNorm(G) for that group.

If there are no compact group actions at all, then $\mathcal{R}(M) \approx$ BDiff(M).

Notice that it is quite easy for us to arrange doppelgangers with the first property. Indeed, the ones we used to ensure a lower bound on volume already have this property. We will soon show how to arrange for the last condition to hold.

Many nonpositively curved manifolds are examples where the second condition holds. This follows from the theorem at the end of the previous section. However, it is unknown whether or not the torus has this property. For example, if the three-dimensional Poincaré conjecture were false, one could construct a nonlinearizable T^n action on T^{n+3}. (There would be a similar phenomenon if one could find a nonstandard smooth structure on T^4.)

Proposition   *Every M has doppelgangers with no symmetry.*

We use the criterion from the end of the previous section. We produce a doppelganger such that the fundamental group has (if the manifold is not M) no finite group of outer automorphisms that could be arising from a group action. Using the theory of homology spheres discussed in chapter 2, we see that there is no difficulty arranging that the homological conditions hold as well.

The way to eliminate outer automorphisms is again to take an amalgamated free product with a knot group as before. We can arrange for the knot complements we use to represent generators of the relative group homology (relative to their boundaries), which then gives the indivisibility of the fundamental class of the doppelganger manifold (assuming it not M).

Before proceeding to discuss theorem 5, it seems worth exploring in a bit more detail what we can learn about filling functions. The answer is, unfortunately, not much, without some additional assumptions. The curves we found that, while nullhomotopic, only bound big disks (or even cycles, if one wants a homological filling function) are examples of "monodromic loops." We will only explain this in the nonsymmmetric case:

Definition.   Let M be a manifold with no compact symmetry groups. We say that a chain C in $\mathcal{R}(M)$ is "nonmonodromic" if the M fibration over C is homotopy equivalent to $C \times M$. (Recall that the cycle sweeps through a set of M's, up to isometry; the lack of symmetry implies that they can be fitted together in a unique well-defined fashion; in the case of $C = S^1$, the monodromy is determined by the unique homotopy equivalence obtained by starting at the base point and comparing nearby fibers using the "neoclassical comparison geometry" and going all the way around the circle to get a self-homotopy equivalence of the base fiber.)

It is quite easy to apply our methods to get information about the nonmonodromic filling functions for nonmonodromic homology cycles. One uses varying seed doppelganger groups, with varying group homology—together with Novikov conjecture results—and then $H\rho$ ideas and/or concordance space theory produce many such cycles with quite different structure. One can arrange for different torsion phenomena, different filling functions, different ranks in different places, and so on.

Proposition   *One can construct regions in $\mathcal{R}(M)$ where the nonmonodromic filling function for $H_k(\ ; \mathbb{Z}_p)$ is specified for each prime separately according to any c.e. sequence of stopping times of Turing machines.*

The proof is straightforward using the material we have developed. The idea is simply to build doppelgangers whose fundamental group, if not that of M, has all sorts of homology sequences built in, and for which the extra part of the fundamental group satisfies the isomorphism conjectures. In that case, in the concordance stable range, one has very powerful control on the diffeomorphisms that are (given) homotopies to the identity.

Unfortunately, it is hard to translate this into an "antifractal" statement that asserts that there are many different regions in $\mathcal{R}(M)$ with genuinely different geometries, because in $\mathcal{R}(M)$, the monodromic and nonmonodromic directions can well interact. Nonetheless, I have no doubt that this is the case quite generally.

One case where this is very likely is where M is antisimple, so that the theory of higher rho invariants can be applied, selectively, in different regions to both construct and detect differing amounts of cohomology. However, even here one looks stuck, because of the autohomotopy equivalences.

Although I do not have a general proof, here is an example that shows that sometimes one can prove that there are many different sorts of local geometries that differ from one another.

Theorem *There are simply connected manifolds with the following property: For any computable f, and all sufficiently large D, there are exponentially in D many points whose balls of radius D (associated with points where the diameter is at most D) in $\mathcal{P}R(M)$ that cannot be f-distortedly matched up with each other. In other words, there is no homeomorphism (or even homotopy equivalence) of a neighborhood of π to one of d which does not distort distances by more than f. (Thus the image of $B(p, D)$ lies within $B(q, f(D))$. Recall that $\mathcal{P}R(M)$ is the pointed version of $\mathcal{R}(M)$.*

If K is a constant, then two balls can be K-distortedly matched up if there are maps with Lipschitz constant K from one ball to the KD-neighborhood of the other, such that on balls of size $D/2$, they satisfy the conditions of homotopy equivalence. (Thus, nearby large balls in a metric space can be matched up with each other for K close to 1.) For a more general function, we are simply allowing the K to increase with D.

The idea is quite simple. We will consider a sequence of groups, which are each either trivial, or have as their $K(\pi, 1)$'s complexes that are homotopy wedges of S^2's after "plusing." We will thicken these complexes up in higher-dimensional Euclidean space, and then take the boundary of their regular neighborhoods. One can readily show that $\pi_i(\text{Aut}(\text{Doppelgangers})) = 0$ for $1 < i < n/2$. Note also that these manifolds as antisimple, so one can find the desired invariants to detect extra elements.

Note that we make genuine use of the pointedness of the moduli space in two ways: first to remove any singularities coming from symmetry, and second to find the fundamental group of the doppelganger: the analysis shows that $\pi_1 \mathcal{P}R(M')$ has a surjection to $\pi_1(M')$ with polycyclic kernel. One can eliminate the second application by using groups whose low skeleton has given plus construction, but which differ in higher skeleta, or by using groups with torsion, where the elements of finite order have homologically different centralizers, so that the isomorphism conjecture will cause them to have rather different homology for their $\mathcal{P}R(M$'s).

We shall not try to squeeze out any more out of these kinds of arguments, since it is clear that we are missing one or two ideas that are preventing us from doing a decent job. Instead, we now move on to the proof of theorem 5.

Proof of Theorem 5. The computability of the filling functions follows from the fact that these manifolds do not have any positive dimension groups of symmetries, and that in the concordance stable range one can actually compute the rational homotopy of their BDiff's. What is more interesting is the non-computablity of the filling function for S^n. The reason for this is the symmetry. Consider circle actions on S^n with fixed set a S^{n-2}, where it is hard to tell Σ from the standard sphere. These produce strata, which if we use doppelgangers that have no circle action, define maximal strata. These circle actions give maps $BS^1 \to \mathrm{BDiff}(S^n)$, which concordance space theory implies are all homotopic, at least restricted to a skeleton in the concordance stable range. For a pair of strata, one thus gets a map $\Sigma\mathbb{CP}^k \to \mathcal{R}(S^n)$. (The symmetric metrics give ways of extending the maps $BS^1 \to \mathrm{BDiff}(S^n)$ over $c\mathbb{CP}^k \to \mathcal{R}(S^n)$; combining two gives a map of $\Sigma\mathbb{CP}^k$ to $\mathcal{R}(S^n)$.)

This map is not injective on homology because there are nonequivariant diffeomorphisms in the normalizer of these circle actions; that is, they extend to $O(2)$ actions. This affects H_3, but not H_5. A computable bound on the filling function for these rational homology groups would give an algorithm to decide which of the Σ are the standard sphere.

4.8 FURTHER DIRECTIONS

In this book we tried to develop a theme: positive application of algorithmic indistinguishability or inseparability of a space from a class of spaces with widely different properties. We have given a number of applications to different variational problems and some applications to the geometry of certain moduli spaces.

I believe that this subject is still in its infancy; the methods and the results suggest many questions, some of which were scattered throughout, some of which I would like to mention here.

Improved Fractality

The definition of "scale" and our understanding of the scales for which it's true that the $f(D)$ deep local minima are $g(D)$ dense, with f and g of the same scale, is quite primitive. It would be much nicer if there were a genuine linear connection between depth and density, although that seems to me quite unlikely.

Different Curvature Conditions

In the final chapter we studied functionals on the space of metrics with two-sided curvature bounds. However, there are geometric precompactness theorems with only a lower bound. It would be good to improve the results to handle this. The main obstacle is the effectiveness of the precompactness. Moreover, in recent years, there has been a great jump in our knowledge of the possible degenerations of manifolds, assuming only a lower bound on the Ricci curvature. Again, this suggests that with this much weaker bound, it might be possible to obtain analogous results. Of course, one's ultimate goal in this direction would be to obtain new Einstein metrics. It is important to realize that this program could produce only metrics with singularities, and the singularities must be sufficiently bad that regularity theory for the Einstein equation cannot kick in.

In the course of this program it would also be useful to deal with Vol in place of diameter as the functional. This seems quite likely to be possible, especially with a lower bound on curvature in view of the well-developed Cheeger-Fukaya-Gromov collapsing theory, and the analyses done on Alexandrov spaces with lower curvature bounds. However, there remain conceptual difficulties in carrying this out, even in the case of two-sided bounds on curvature.

Antifractality

In the very last section we described some results which show that a naive universal self-similarity cannot be present in $PR(M)$, for at least certain M. No doubt this is true for (almost) all M, even in the unpointed case. The main issue is technical: it is hard to control the contribution of self-homotopy equivalences of doppelgangers to the associated regions of moduli space. I can imagine several ways around this, and would hope that one can give very large lower bounds on the number of different geometries one sees on balls as one moves around moduli space. One approach to this could make use of improvements of the logical and homological group theory discussed in chapter 1 from finitely presented groups to ones with finite skeleta. Here are two concrete questions, whose positive solution could be quite helpful:

Q1.   Is it true that for every finite simply connected complex there is a sequence of finite complexes which are either homotopy equivalent to this given complex, or if not, are aspherical? (As always, the two situations should not be algorithmically distinguishable.)

Q2. Is there a universal group (i.e., a group which contains all finitely presented groups) which itself has finite skeleta? A related issue: if one wants to obtain a given (computably presented abelian) group as H_n, can this be done keeping the $n - 1$ skeleton finite, and presumably with vanishing homology below dimension n?

In any case, the basic question is to prove that there are an exponentially large number of different types of basins, that have quite different homotopy theory even if one allows very distorting isomorphisms among them.

Dimension Four

For certain 4-manifolds we have shown the existence of infinitely many local minima. However there are many things we do not know. Probably the most pressing are

Q1. Is S^4 algorithmically recognizable? (This is related to the question of whether there is an algorithm to decide whether an n-generator n-relator group is trivial.)

Q2. Can one realize all elements of $H_4(B\pi)$ that lie in the image of the Hurewicz homomorphism by homology spheres? This seems quite unlikely, but I don't know how to rule it out.

In dimension three, geometricization à la Thurston implies that the fractal picture is entirely missing, except perhaps at certain small scales.

Precise Connection to Kolmogorov Complexity

Nabutovsky and I have conjectured the following:

Let M be a closed smooth manifold of dimension $n > 3$. For any Riemannian metric g on M with $|K| \leq 1$, consider the set

$$S(D) = \{\text{integers with Kolmogorov complexity} \leq \text{Diam}(g)\}.$$

Then for any integer in $S(\text{const}(M)(D^n + 1))$ there are more than $\exp(D^n)$ local minima for diameters of depth between N/c and $\exp\exp(N + D)$ that are within $\exp\exp(N + D)$ of g in the path metric.

(It is conceivable that even a single exponential might suffice. Of course, exp here just means some exponential with base > 1, not necessarily $2.718\ldots$)

The fact that we could encode all Turing machines lends support to the idea that all depths of small complexity can arise. And, as is well known (exercise!), the largest number with a small complexity grows faster than any computable function; this is a restatement of our theorem about noncomputably deep local minima.

Quantum Gravity

In some models of quantum gravity, one does integration (numerical or otherwise, and probably with a renormalization, but let's ignore that point) over spaces like $\mathcal{R}(M)$. A fundamental problem is that we do not know its metric entropy: How many balls does it take to cover it? The physicists believe that it is exponential, but what is obvious to mathematicians is $\exp(V \log V)$.

Nabutovsky has suggested that the encoding of logical problems into $\mathcal{R}(M)$ leads to the possibility of doing experiments to discover the answers to unsolvable problems. Another way to say this is that "quantum gravity analog computers" could conceivably compute more than any digital computer. (We are ignoring time issues here!)

This is different from the current interest in quantum computers, which compute exactly the same functions as ordinary computers do, but just do it faster.

To give a toy version of this idea, let us consider a type of function for which convergence issues do not arise, and let us assume, in a science fiction sort of way, that one can design an experiment where very careful measurements will actually find for us more and more digits of this function on an input.

Let $\theta(M, t) = \Sigma\Delta(n, M)(\exp(-n^2 t))$, where $\Delta(n, M)$ is the number of triangulations of M with at most n simplices (and perhaps a bound on the local geometry, if you like). $\theta(M, t)$ converges for positive t.

(Ironic) Proposition

(Left to the Reader) For any $t > 0$ a computable real number, $\theta(M, t)$ is a finite to 1 invariant of PL manifolds. However, for no manifold of dimension > 4, and for many 4-manifolds, it is not computable; in particular it is transcendental over $\mathbb{Q}[\exp(\text{algebraics})]$, π, $\log 13203$, etc.].

The interesting thing is now the following: Machines that can refer to calculations of $\theta(M, t)$'s can compute exactly the same things as ones which can access $\theta(M, 1)$, for $S^2 \times S^2 \# S^2 \times S^2 \# \cdots \# S^2 \times S^2$ enough times (or any 5-manifold). Moreover, these are the same as the functions that can be computed by an ordinary computer in infinite time.

These last functions have a computer working to compute them; the machine can change its mind finitely often (otherwise the answer is "infinity" or "indeterminate"); however, at no finite stage does the computer let us know whether or not it is done.

Whether quantum gravity computers will become a useful technology only time will tell. (Perhaps it will take an infinite amount of time!)

Other Variational Problems

Overall, the scheme that was applied here (except in the last sections of the last part) is quite general. Any problem with a sufficient logical or computational complexity is apt to have many solutions.

In geometry we have seen that this often follows from diffeomorphism invariance, although we have seen a few such problems with other sources of logical complexity.

It would be reasonable to believe that there are also some nongeometric examples (if there is *anything* at all which is truly *nongeometric*) where similar phenomena arise.

NOTES

That Riem/Diff is at all a nice space is basically due to Ebin. (See also Bourginon.) Some more analysis of its homotopy type and geometry near the symmetric metrics has been provided by Maher, in unpublished work.

D. Ebin. *The manifold of Riemannian metrics*. In *Global Analysis*. Proceedings of the Symposium on Pure Mathematics, vol. XV, Berkeley, Calif., 1968, pp. 11–40. American Mathematical Society, Providence, R.I., 1970.

J. P. Bourguignon. *Une stratification de l'espace des structures riemanniennes* (French). Compositio Math. 30 (1975), 1–41.

Maher, in particular, discusses the bearing of the geometricization of 3-manifolds for the homotopy type of Riem/Diff of 3-manifolds. For a general discussion of the geometricization conjecture, see

P. Scott. *The geometries of 3-manifolds*. Bull. London Math. Soc. 15 (1983), 401–487.

In higher dimensions, a great focus has been the search for Einstein metrics, that is, metrics whose Ricci curvature (a contraction of the usual Riemannian curvature) is proportional to the metric (they are both symmetric 2-tensors). It has been suggested that these could be the "best metrics" a manifold might have.

Of course, the philosophy we embrace suggests that widespread existence should imply widespread nonuniqueness, and that perhaps these metrics will not actually turn out to be so useful.

For a very useful general introduction to the theory of Einstein manifolds, see the book "by" Besse:

A. Besse. *Einstein Manifolds*. Ergebnisse der Mathematik und ihrer Grenzgebiete (3) 10. Springer-Verlag, Berlin, 1987.

This book in particular explains the variational significance of Einstein metrics. See also the papers of Mike Anderson on some ideas about carrying through the variational approach to constructing such metrics.

M. Anderson. *Degeneration of metrics with bounded curvature and applications to critical metrics of Riemannian functionals*. In *Differential Geometry: Riemannian Geometry* Proceedings of the Symposium on Pure Mathematics, vol. 54, Los Angeles, Calif., 1990, part 3, 53–79. American Mathematical Society, Providence, R.I., 1993.

In dimension four, there are numerous results now, both from an entropy point of view and using the Seiberg-Witten theory, that show nonexistence and uniqueness of smooth solutions to the Einstein equations. See

G. Besson, G. Courtois, and S. Gallot. *Les variétés hyperboliques sont des minima locaux de l'entropie topologique*. Invent. Math. 117 (1994), no. 3, 403–445.

C. LeBrun. *On four-dimensional Einstein manifolds*. In *The Geometric Universe* (Oxford, 1996). 109–121. Oxford University Press, Oxford, 1998.

For these manifolds, the ideas we sketched actually do suffice to produce some of the geometric complexity of $\mathcal{R}(M)$ that we established in high dimensions. However, it seems quite conceivable that there are extremal metrics that do not have sufficient smoothness for the application of these analytic techniques.

For some more discussion of the issues involved in the analysis of Einstein metrics, see

A. Nabutovsky. *Einstein structures: Existence versus uniqueness*. Geom. Funct. Anal. 5 (1995), no. 1, 76–91.

The main results of this chapter are mainly taken from

A. Nabutovsky and S. Weinberger. *Variational problems for Riemannian functionals and arithmetic groups*. Publ. Math. d'IHES 92 (2001), 5–62.

_____. *The fractal geometry of Riem/Diff I*. Geometria Dedicata 101 (2003), 1–54.

Comparison geometry did not start with Cheeger's thesis; for example, the Rauch comparison theorem preceded this. However, the nature of the subject changed with the finiteness theorem. There are now many references and surveys of the ideas we discussed in section 4.2.

J. Cheeger. *Finiteness theorems for Riemannian manifolds*. Amer. J. Math. 92 (1970), 61–74.

M. Gromov. *Metric Structures for Riemannian and Non-Riemannian Spaces*. Based on the 1981 French original. With appendixes by M. Katz, P. Pansu, and S. Semmes. Translated from the French by Sean Michael Bates. Progress in Mathematics 152. Birkhäuser, Boston, Mass., 1999.

P. Peterson. *Riemannian Geometry*. Graduate Texts in Mathematics 171. Springer-Verlag, New York, 1998.

I should point out that the smooth version of Cheeger's finiteness theorem was asserted in

J. Cheeger and D. Ebin. *Comparison Theorems in Riemannian Geometry*. North-Holland Mathematical Library 9. North-Holland, Amsterdam/American Elsevier, New York, 1975,

another fine influential and valuable, if somewhat dated, introduction to comparison geometry. The first published proof is due to Peters:

S. Peters. *Cheeger's finiteness theorem for diffeomorphism classes of Riemannian manifolds*. J. Reine Angew. Math. 349 (1984), 77–82.

Explicit bounds on the number of diffeomorphism types of manifolds in this class were worked out by

T. Yamaguchi. *On the number of diffeomorphism classes in a certain class of Riemannian manifolds.* Nagoya Math. J. 97 (1985), 173–192.

Gromov's book is a much expanded English second edition of Gromov's classic which introduced Gromov-Hausdorff space and convergence methods. That book also emphasizes lower Ricci curvature bounds, which will indeed be extremely relevant to the problem of Einstein metrics.

In particular, the simplicial norm controls volume merely in the presence of a lower Ricci bound.

I believe that the first general smoothing theorem was hidden in

J. Cheeger and M. Gromov. *On the characteristic numbers of complete manifolds of bounded curvature and finite volume.* In *Differential Geometry and Complex Analysis*, 115–154. Springer, Berlin, 1985.

These results were improved by analytic means by U. Abresch. See the papers

J. Bemelmans, Min-Oo, and E. Ruh. *Smoothing Riemannian metrics.* Math. Z. 188 (1984), no. 1, 69–74.

S. Bando. *Real analyticity of solutions of Hamilton's equation.* Math. Z. 195 (1987), no. 1, 93–97.

See also

P. Petersen, G. Wei, and R. Ye. *Controlled geometry via smoothing.* Comment. Math. Helv. 74 (1999), no. 3, 345–363,

which has an elegant reexplication of these ideas with some new applications.

The results of section 3 are all taken from the first Nabutosky-Weinberger paper mentioned above, except for the Kolmogorov complexity method of getting c^{c^d} local minima, which is a slight modification of

A. Nabutovsky. *Geometry of the space of triangulations of a compact manifold.* Commun. Math. Phys. 181 (1996), no. 2, 303–330

that Nabutovsky and I wrote out in unpublished work.

The following is the main result of Nabutovsky's paper.

Theorem   *For any $n > 4$, there is a constant $C(n)$ such that for any compact n-manifold, and any computable $t(N)$, for all sufficiently large N, there are at least $C(n)^N$ triangulations of M with at most N simplices, such that the distance between any two of the triangulations is at least $t(N)$. Moreover, these triangulations can be chosen minimal with respect to any "simplification algorithm" (i.e., one which minimizes a computable aspect of a triangulation by comparing to "nearby neighbors" of at most a given computable distance away).*

Note that $c(n)$ is independent of M, $t(N)$, the "simplification algorithm," and so on, but the lower bound on N required to make the theorem true will depend on these choices.

Gromov's theorem bounding volume using the simplicial norm appears in his paper

M. Gromov. *Volume and bounded cohomology*. Inst. Hautes Études Sci. Publ. Math. 56 (1982), 5–99 (1983).

As noted in our paper, one can also apply the results of

M. Gromov. *Filling Riemannian manifolds*. J. Diff. Geom. 18 (1983), no. 1, 1–147

to give an alternative approach; doing this requires more complicated ad hoc constructions, but then avoids the very deep results of Clozel (and others) mentioned in chapter 1, but instead depends on deep results of (at least) A. Borel, N. Wallach, and others.

The $C^{1,\alpha}$ structure on limit points is often called the Cheeger-Gromov compactness theorem. Proofs based on Jost and Karcher's harmonic coordinates appear in

S. Peters. *Convergence of Riemannian manifolds*. Compositio Math. 62 (1987), no. 1, 3–16.

R. Greene and H. Wu. *Lipschitz convergence of Riemannian manifolds*. Pacific J. Math. 131 (1988), no. 1, 119–141.

A synthetic approach within the context of Alexandrov geometry can be found in the work of Nikolaev and Berestovskii.

V. N. Berestovskii. *Introduction of a Riemannian structure in certain metric spaces* (Russian). Sibirsk. Mat. Z. 16 (1975), no. 4, 651–662.

I. Nikolaev. *Smoothness of the metric of spaces with bilaterally bounded curvature in the sense of A. D. Aleksandrov* (Russian). Sibirsk. Mat. Zh. 24 (1983), no. 2, 114–132.

The results of section 4 are somewhat more precise than, but essentially follow the same ideas as, the second Nabutosky-Weinberger paper. Note that the version of scales used here is more precise than the one used in that paper.

The "fractality" in the current version is quite easy: clearly, the large basins must have many small basins coming off them, and these smaller basins must have yet smaller scale ones coming off those, and so on.

This qualitative picture with the old version of scale is more difficult to interpret. Early on, we used a Turing degree definition, which led only to an "infinitely many d" result. With time, we were led to use versions more directly related to c.e. sets, and the relevant existence of "almost all d" results regarding the layering of different scales was produced on request by Bob Soare.

While we no longer absolutely require it, we would like to express gratitude for his construction which confirmed for us our basic mental image. That one can be led to the same general qualitative result, by being more refined in one's understanding either of the dynamics of the activities of particular Turing machines or of the interaction between the metric entropy of moduli spaces and the shapes of doppelgangers, further strengthens (at least in my mind) the connection between computation and moduli.

The results of section 4.4 are mainly refinements of some of those of the second Nabutovsky-Weinberger paper mentioned above.

The material in section 4.5 is all well known. Smoothing theory was first developed in the context of smoothing PL manifolds. The canonical reference is

M. Hirsch and B. Mazur. *Smoothings of Piecewise Linear Manifolds*. Annals of Mathematics Studies 80. Princeton University Press, Princeton, N.J. University of Tokyo Press, Tokyo, 1974.

For the smoothing theory of topological manifolds, one requires the deeper

R. Kirby and L. Siebenmann. *Foundational Essays on Topological Manifolds, Smoothings, and Triangulations*. With notes by John Milnor and Michael Atiyah. Annals of Mathematics Studies, No. 88. Princeton University Press, Princeton, N.J. University of Tokyo Press, Tokyo, 1977.

(This book also explains many other ideas related to surgery theory, simplicial methods, and the like. While it is not easy, it is rewarding reading.)

The space BD, its relation to block bundles, the L-spaces, and their methods of analysis are due to Quinn and were developed in his thesis. (An earlier, more special, version appeared in work of Casson.)

F. Quinn. *A geometric formulation of surgery*. In *Topology of Manifolds*. Proceedings of the Institute, University of Georgia, Athens, Ga., 1969, pp. 500–511. Markham, Chicago, Ill, 1970.

See also

D. Burghelea, R. Lashof, and M. Rothenberg. *Groups of Automorphisms of Manifolds*. With an appendix ("The topological category") by E. Pedersen. Lecture Notes in Mathematics 473. Springer-Verlag, Berlin, 1975.

S. Weinberger. *The Topological Classification of Stratified Spaces*. University of Chicago Press, Chicago, Ill., 1994.

In order to get information about isotopy classes of diffeomorphisms, one first uses surgery theory to get the pseudoisotopy information, and then one needs to analyze the difference between pseudoisotopy and isotopy. The first breakthrough result is due to Cerf:

J. Cerf. *La stratification naturelle des espaces de fonctions différentiables réelles et le théorème de la pseudo-isotopie*. Inst. Hautes Études Sci. Publ. Math. 39 (1970), 5–173,

who showed that for high-dimensional simply connected manifolds, pseudoisotopic diffeomorphisms are isotopic. The analysis for nonsimply connected manifolds was given in the pair of papers by A. Hatcher and J. Wagoner,

A. Hatcher and J. Wagoner. *Pseudo-Isotopies of Compact Manifolds*. With English and

French prefaces. Astérisque no. 6. Société Mathématique de France, Paris, 1973,

A. Hatcher. *The second obstruction for pseudo-isotopies.* Bull. Amer. Math. Soc. 78 (1972), 1005–1008,

with a correction by Igusa:

K. Igusa. *What happens to Hatcher and Wagoner's formulas for $\pi_0 C(M)$ when the first Postnikov invariant of M is nontrivial?* In *Algebraic K-Theory, Number Theory, Geometry and Analysis* (Bielefeld, 1982), 104–172. Lecture Notes in Mathematics 1046. Springer-Verlag, Berlin, 1984.

Hatcher initiated higher concordance space theory, but the definitive treatments of it are due to

F. Waldhausen. *Algebraic K-theory of topological spaces. I.* In *Algebraic and Geometric Topology.* Proceedings of the Symposium on Pure Mathematics, Stanford University, vol. XXXII, Stanford, Calif., 1976, part 1, 35–60. American Mathematical Society, Providence, R.I., 1978

(and a number of later papers) and

K. Igusa. *The stability theorem for smooth pseudoisotopies.* K-theory 2 (1988), no. 1–2, vi+355 pp.

for the concordance stable range results. A very useful survey that explains the statements of these and many other results, as well as their own work on the marriage between pseudoisotopy and surgery theories, is

M. Weiss and B. Williams. *Automorphisms of Manifolds.* Surveys in surgery in honor of C.T.C. Wall, vol. 2, 165–220. Princeton University Press, Princeton, 2001.

They proved, in particular, the result that torus stabilization kills the difference between BD and BDiff. Their work also explains in much more precise terms the final result mentioned in this section connecting the invariant part of the concordance space theory and diffeomorphism groups. An earlier form of this, as a spectral sequence, is explained in

A. Hatcher. *Concordance spaces, higher simple homotopy theory, and applications,* In *Algebraic and Geometric Topology,* Proceedings of the Symposium on Pure Mathematics, vol. XXXII, part 1, Stanford University, Stanford, Calif., 1976, American Mathematical Society, Providence, R.I., 1978,

and a rational version of how to calculate can be found in

D. Burghelea and Z. Fiedorewicz. *Hermitian Algebraic K-theory of simplicial rings and topologcial spaces.* J. Math. Pures Appl. 64 (1985), 175–235.

To actually do, say, rational calculations, one would start with the (easy) isomorphism

$$\mathbf{A}(B\pi) \cong \mathbf{K}(\mathbb{Z}\pi) \oplus \mathbb{Q}$$

with methods for analyzing the relative space $A(X, B\pi)$. The valuable tool for this is the Goodwillie calculus. A suitable general introduction is Goodwillie's ICM talk:

T. Goodwillie. *The differential calculus of homotopy functors*. In *Proceedings of the International Congress of Mathematicians* vols. I and II, Kyoto, 1990, 621–630. Mathematical Society of Japan, Tokyo, 1991

and also Weiss's expository article that adapts the theory to another setting,

M. Weiss. *Calculus of embeddings*. Bull. Amer. Math. Soc. 33 (1996), 177–187.

The isomorphism conjectures have an intricate history. The analogue in the theory of C^*-algebras is the Baum-Connes conjecture:

P. Baum and A. Connes. *Geometric K-theory for Lie groups and foliations*. Enseign. Math. (2) 46 (2000), no. 1–2, 3–42.

(This is a publication of an approximately twenty-year-old manuscript!) Versions in topology were suggested in the work of Quinn on algebraic K-theory.

F. Quinn. *Algebraic K-theory of poly-(finite or cyclic) groups*. Bull. Amer. Math. Soc. (N.S.) 12 (1985), no. 2, 221–226.

_____. *Applications of topology with control*. Proceedings of the International Congress of Mathematicians vols. 1 and 2 (Berkeley, Calif., 1986), 598–606. American Mathematical Society, Providence, R.I., 1987.

These were followed by many others. Chapter 13 of my book (referred to above) explains how heuristic reasoning involving the Borel conjecture and very general aspects of the theory of stratified spaces leads one directly in this direction. However, the appearence of Nil terms in the Bass-Heller-Swan formula in algebraic K-theory, and the work of Waldhausen and Cappell on amalgamated free products in K-theory and L-theory, respectively, indicated that things had to be much more complicated in some mysterious way.

Farrell and Jones, led by their deep positive results and methods of proof of the Borel conjecture for hyperbolic and related manifolds, took the bull by its horns and conjectured that all one has to do is replace finite groups by virtually cyclic ones in the formulation of such conjectures. As far as we know, this simple modification is all that is necessary. See

T. Farrell and L. Jones. *Isomorphism conjectures in algebraic K-theory*. J. Amer. Math. Soc. 6 (1993), no. 2, 249–297.

Very recently, F. Connolly and J. Davis, *The surgery obstruction groups of the infinite dihedral group* (http://front.math.ucdavis.edu/math.GT/0306054), and also M. Banagl and A. Ranicki (http://front.math.ucdavis.edu/math.AT/0304362), have computed the UNil groups occurring in $\mathbf{L}_n(\mathbb{Z}_2 * \mathbb{Z}_2)$.

The results of simply connected surgery are due to W. Browder and S. Novikov, although the formulation I gave is due to D. Sullivan. The situation for finite fundamental group is a synthesis of these results with work of Wall with Atiyah, Patodi, and Singer.

W. Browder. *Surgery on simply-connected manifolds*. Ergebnisse der Mathematik und ihrer Grenzgebiete 65. Springer-Verlag, New York, 1972.

S. Novikov. *Analogues hermitiens de la K -théorie. Actes du Congrès International des Mathématiciens* (Nice, 1970) vol. 2, 39–45. Gathier-Villars, Paris, 1971.

C.T.C. Wall. *Surgery on Compact Manifolds.* London Mathematical Society Monographs, no. 1. Academic, London, 1970.

M. Atiyah, V. Patodi, and I. Singer. *Spectral asymmetry and Riemannian geometry: I.* Math. Proc. Cambridge Philos. Soc. 77 (1975), 43–69; *II.* Math. Proc. Cambridge Philos. Soc. 78 (1975), no. 3, 405–432; *III.* Math. Proc. Cambridge Philos. Soc. 79 (1976), no. 1, 71–99.

Antisimple manifolds were introduced and studied in a very pretty unpublished manuscript of Jean Claude Hausmann. He did publish the theory for the special case of knots:

J. Cl. Hausmann. *Noeuds antisimples* (French). In *Knot Theory.* Proceedings of the Seminar, Plans-sur-Bex, Switzerland, 1977, 171–202. Lecture Notes in Mathematics, 685. Springer-Verlag, Berlin, 1978.

I applied them to the study of secondary higher rho invariants in my paper

S. Weinberger. *Higher ρ-invariants.* In *Tel Aviv Topology Conference: Rothenberg Festschrift* (1998), 315–320. Contempory Mathematics 231. American Mathematical Society, Providence, R.I., 1999.

This work was motivated by an earlier paper of

J. Lott. *Higher η-invariants. K* -theory 6 (1992), no. 3, 191–233,

which works for a smaller class of groups, but for more general elliptic operators. There have since been a number of other papers on this general topic; a very nice extension to some non-anti-simple situations is

E. Leichtenam, W. Lück, and M. Kreck. *On the cut and paste property for higher signatures.* Topology 41 (2002), 725–744.

The papers of Mishchenko and Ranicki that my work depends on are

A. Miscenko. *Homotopy invariants of multiply connected manifolds: III: Higher signatures* (Russian). Izv. Akad. Nauk SSSR Ser. Mat. 35 (1971), 1316–1355.

A. Ranikci. *The algebraic theory of surgery. I. Foundations.* Proc. London Math. Soc. (3) 40 (1980), no. 1, 87–192.

The reader should be warned that Mischenko's paper is seriously flawed, but not in defining the symmetric signature. The results about symmetric signatures of total spaces are implicit in at least

S. Cappell and S. Weinberger. *Replacement of fixed sets and of their normal representations in transformation groups of manifolds.* In *Prospects in Topology* (Princeton, N.J., 1994), 67–109. Annals of Mathematical Studies 138. Princeton University Press, Princeton, N.J., 1995.

W. Lueck and A. Ranicki. *Surgery obstructions of fibre bundles.* J. Pure Appl. Algebra 81 (1992), no. 2, 139–189.

I should also mention here some papers that apply techniques related to the Novikov/ Borel conjectures to study the original Atiyah-Patodi-Singer invariant:

W. Neumann. *Signature related invariants of manifolds.   I, Monodromy and γ-invariants.* Topology 18 (1979), 2147–2172.

S. Weinberger. *Homotopy invariance of η-invariants.* Proc. Natl. Acad. Sci. U.S.A. 85 (1988), no. 15, 5362–5363.

M. Farber and J. Levine. *Jumps of the η-invariant.* With an appendix by Shmuel Weinberger (Rationality of ρ-invariants). Math. Z. 223 (1996), no. 2, 197–246.

N. Keswani. *Relative eta-invariants and C^*-algebra K-theory.* Topology 39 (2000), no. 5, 957–983.

The paper of Cheeger and Gromov cited above introduces a von Neumann algebraic invariant that modifies the Atiyah-Patodi-Singer invariant in a different direction for infinite fundamental group. This extension of ρ-invariants behaves rather differently from the Hρ theory. A partial analysis of its homotopy invariance properties was initiated in the papers

V. Mathai. *L^2-invariants of covering spaces.* In *Geometric Analysis and Lie Theory in Mathematics and Physics*, 209–242. Australian Mathematical Society Lecture Series 11, Cambridge University Press, Cambridge, 1998.

N. Keswani. *Von Neumann η-invariants and C^*-algebra K-theory.* J. London Math. Soc. (2) 62 (2000), no. 3, 771–783.

S. Chang and S. Weinberger. *On invariants of Hirzebruch and Cheeger-Gromov.* Geometry and Topology 7 (2003), 311–319.

The study of group actions (i.e., compact groups of symmetries of manifolds) is one of my favorite sub-branches of topology. It has gone through many eras, each introducing different techniques and uncovering new phenomena. Good general references, each from a different era, are

A. Borel. *Seminar on Transformation Groups.* With contributions by G. Bredon, E. E. Floyd, D. Montgomery, and R. Palais. Annals of Mathematics Studies no. 46. Princeton University Press, Princeton, N.J. 1960; *Proceedings of the Second Conference on Compact Transformation Groups* (University of Massachusetts, Amherst, Mass., 1971), Parts I and II. Edited by H. T. Ku, L. N. Mann, J. L. Sicks, and J. C. Su. With an introduction by Deane Montgomery. Lecture Notes in Mathematics 298 and 299. Springer-Verlag, Berlin, 1972.

P. Connor and E. Floyd. *Differentiable Periodic Maps.* Ergebnisse der Mathematik und ihrer Grenzgebiete, N.F., 33. Academic, New York/Springer-Verlag, Berlin, 1964.

W. Browder. *Surgery and transformation groups.* In *Proceedings of the Conference on Transformation Groups* (New Orleans, La., 1967), 1–46. Edited by P. Mostert. Springer-Verlag, New York, 1968.

G. Bredon. *Introduction to Compact Transformation Groups*. Pure and Applied Mathematics vol. 46. Academic Press, New York, 1972. *Group Actions on Manifolds*. Proceedings of the AMS-IMS-SIAM Joint Summer Research Conference (University of Colorado, Boulder, 1983). Edited by Reinhard Schultz. Contemporary Mathematics 36. American Mathematical Society, Providence, R.I., 1985.

The first theorem of this section is entirely standard; as far as I know it was first stated in my paper with Nabutovsky, but it is based on W. Y. Hsiang's thesis:

W. Y. Hsiang. *On the unknottedness of the fixed-point set of differentiable circle group actions on spheres*. P. A. Smith conjecture. Bull. Amer. Math. Soc. 70 (1964), 678–680.

A good reference for this type of construction and how to work modulo it is the beautiful lectures of M. Davis,

M. Davis. *Multiaxial Actions on Manifolds*. Lecture Notes in Mathematics 643. Springer-Verlag, Berlin, 1978.

References for the other methods of modifying actions can be found in the Boulder conference mentioned above, as well as the book

T. Petrie and J. Randall. *Transformation Groups on Manifolds*. Monographs and Textbooks in Pure and Applied Mathematics 82. Marcel Dekker, New York, 1984

for the issues of equivariant tangent bundles. The paper

S. Weinberger. *Nonlinear averaging, embeddings, and group actions*. In *Tel Aviv Topology Conference: Rothenberg Festschrift* (1998), 307–314. Contemporary Mathematics 231. American Mathematical Society, Providence, R.I., 1999

begins a development of a systematic connection between embedding theory and group actions. Item (3) in section 6 is the subject of "converses to Smith theory," a subject initiated by Lowell Jones in his thesis

L. Jones. *The converse to the fixed-point theorem of P. A. Smith: I*. Ann. Math. (2) 94 (1971), 52–68

and developed in many other papers by other authors (such as A. Assadi, W. Browder, S. Cappell, J. Davis, P. Loffler, and S. Weinberger). I have a survey of some applications of it in the Boulder conference, which is, of course, now out of date.

The results on non-p-groups are due, mainly, to R. Oliver (although T. Petrie and T. tom Dieck have also substantially extended the scope of this theory). His first and last papers on this topic are

R. Oliver. *Fixed-point sets of group actions on finite acyclic complexes*. Comment. Math. Helv. 50 (1975), 155–177.

R. Oliver. *Fixed point sets and tangent bundles of actions on disks and Euclidean spaces*. Topology 35 (1996), no. 3, 583–615.

The results that cut down symmetry referred to at the end are largely motivated by an old result of Borel's on group actions on aspherical manifolds. (See, e.g., the papers by

Connor and Raymond in the Amherst conference.) Other papers that extend the original idea of Borel in directions of the sort given are

A. Assadi and D. Burghelea. *Symmetry of manifolds and their lower homotopy groups.* Bull. Soc. Math. France 111 (1983), no. 2, 97–108.

W. Browder and W. C. Hsiang. *G-actions and the fundamental group.* Invent. Math. 65 (1981/82), no. 3, 411–424.

H. Donnelly and R. Schultz. *Compact group actions and maps into aspherical manifolds.* Topology 21 (1982), no. 4, 444–455.

R. Schoen and S. T. Yau. *Compact group actions and the topology of manifolds with nonpositive curvature.* Topology 18 (1979), no. 4, 361–380.

S. Weinberger. *Group actions and higher signatures: II.* Commun. Pure Appl. Math. 40 (1987), no. 2, 179–187.

J. Rosenberg and S. Weinberger. *An equivariant Novikov conjecture.* With an appendix by J. P. May. K-Theory 4 (1990), no. 1, 29–53.

P. Ding. *Semisimple group actions on nonsimply connected manifolds.* Courant Ph.D. thesis, 2003.

I should mention that the group of all lifts of the group action to the universal cover = fundamental group of the nonsingular part (under the quite common "gap hypotheses") was called the orbifold fundamental group by Thurston, and the name seems to have stuck. The relevant facts about group extensions that are applied in our argument can be found in

K. Brown. *Cohomology of Groups.* Corrected reprint of the 1982 original. Graduate Texts in Mathematics 87. Springer-Verlag, New York, 1994.

(There is also some discussion in Gromov's "Volume and bounded cohomology.") Another very useful result which can be used to eliminate symmetry for simply connected manifolds is due to Atiyah and Hirzebruch. It asserts that the A-genus of a compact smooth spin manifold with a positive-dimensional Lie group action vanishes.

M. Atiyah and F. Hirzebruch. *Spin-manifolds and group actions. Essays on Topology and Related Topics* (Mémoires dédiés à Georges de Rham), 18–28. Springer-Verlag, New York 1970.

This theorem implies, for instance, that no product of K-3 (= Kummer) surfaces admits a positive-dimensional Lie group action.

(There is a more subtle vanishing theorem for torsion type invariants in the presence of nonabelian Lie group actions that can be applied to exotic spheres, for example:

B. Lawson and S. T. Yau. *Scalar curvature, non-abelian group actions, and the degree of symmetry of exotic spheres.* Comment. Math. Helv. 49 (1974), 232–244.

The proof of this result depends on using the action to construct metrics of positive scalar curvature.)

Finally, the finiteness theorems that I referred to in section 8 where only a lower bound on curvature is assumed are to be found in

K. Grove and P. Petersen. *Bounding homotopy types by geometry*. Ann. Math. (2) 128 (1988), no. 1, 195–206.

K. Grove, P. Petersen, and Wu. *Geometric finiteness theorems via controlled topology*. Invent. Math. 99 (1990), no. 1, 205–213.

S. Ferry (partly with A. N. Dranishnikov) has developed a very deep topological theory that extends this differential geometric work in several directions. A sample paper in this direction is

S. Ferry. *Topological finiteness theorems for manifolds in Gromov-Hausdorff space*. Duke Math. J. 74 (1994), no. 1, 95–106.

I believe that Ferry's ideas can well be applicable to proving effective precompactness results for some of the other cases of geometric functionals with weaker curvature bounds.

Quantum gravity raises a number of other issues for the circle of ideas that we have been considering. An important one is whether or not the integrations that are made on $\mathcal{R}(M)$ and its relatives are actually only first attempts, and one should really have to rescale in some fashion and take limits, or perhaps renormalize integrations that take place at larger and larger scales. It seems likely that some spaces related to Embeddings(M^n in $\mathbb{R}^{n+1}$) might be more easily analyzed to see whether logic survives renormalization.

Index